Formulas, Ingredients and Production of Cosmetics

Hiroshi Iwata · Kunio Shimada

Formulas, Ingredients and Production of Cosmetics

Technology of Skin- and Hair-Care Products in Japan

Springer

Hiroshi Iwata
ERICA Co., Ltd.
Saitama, Japan

Kunio Shimada
NOF Corporation
Tokyo, Japan

Technical translator
Yuko Makuuchi

ISBN 978-4-431-54060-1 ISBN 978-4-431-54061-8 (eBook)
DOI 10.1007/978-4-431-54061-8
Springer Tokyo Heidelberg New York Dordrecht London

Library of Congress Control Number: 2012947809

Printed on acid-free paper

Springer is part of Springer Science+Business Media (www.springer.com)

Preface

I have long wished to publish not an ordinary book but a really useful one for cosmetics researchers in the world, believing it is my mission as I have long worked in this field. Today, many books are available from which readers can learn the basics of designing cosmetics formulations. However, most of the authors work or have worked for private companies, and thus the books do not contain the knowledge that constitutes companies' secret know-how but only the contents for which the companies or organizations approved disclosure. I doubt whether the information is sufficient for readers to actually design cosmetics.

The coauthor of this book, Mr. Hiroshi Iwata, is an old friend of mine. He is an independent consultant and learned about cosmetics by studying them himself over many years. He has given guidance to major cosmetics companies not only in Japan but also overseas. He can impart knowledge freely without being restricted by any organization. He is highly capable and has abundant knowledge about cosmetic formulation, and there are not many people like him even among highly experienced researchers. I believe Mr. Iwata is the most suitable person for writing this book, and I sincerely thank him for spending much time and effort in spite of his busy schedule.

The translator, Yuko Makuuchi, is an experienced expert in Japanese–English translation of scientific and technical writing.

Dr. Daisaku Ikeda (President of Soka Gakkai International), who is my lifelong teacher and has received more than 300 academic awards from all parts of the world, said during an interview with Dr. Vincent Harding, "What is the real purpose of science? It is the happiness of the human being." I believe that development of superior new cosmetics products as a result of hard, accumulated research work has led and will lead to the happiness of people. If this book helps people engaged in research and development of cosmetics formulations, those who work for cosmetics companies, and students studying cosmetics, it will be a great joy for me as an editor.

Finally, I thank Dr. Shinichi Koizumi of Springer Japan KK for giving us an opportunity to publish this book.

Tokyo, Japan Kunio Shimada

Preface

Japan is one of the three largest cosmetics markets in the world. It holds a unique place in the global industry as it has its own set of regulations and technological strengths. Higher safety standards, among other factors, have had the positive impact of forcing more research and thus pushing the science of cosmetics to a very high level. Japan's cosmetics industry continues to focus on fundamental research, relying on cosmetics companies themselves to utilize their superior know-how to bring innovation and refinement to cosmetics and cosmetics-related technologies.

Today, young cosmetics researchers who have completed their graduate studies and have entered a cosmetics company are put through several years of training before they become qualified to design cosmetics formulations themselves. They are trained so that they can design formulas not by a process of logic but by heart, like craftsmen, chefs, or carpenters. To the authors of this book, this kind of training seems a terrible waste of labor and time. To address this issue and allow young cosmetics scientists to design novel formulations and effectively bring greater diversity of innovation to the industry, this book provides a key set of skills and the knowledge necessary for those so inspired.

The book will help readers prepare formulas at the ordinary commercial product level and then eventually design their own original, and possibly high-level, formulations. This volume is designed to provide the comprehensive knowledge and instruction to researchers, that is, the know-how necessary for designing cosmetic products.

The book's chapters cover a comprehensive list of topics and are divided as follows: Chap. 1 is on the basics of cosmetics, including ingredients and dosage. Chapter 2 describes the raw materials of cosmetics and their application, that is, how ingredients like oils, fatty acids, macromolecules, and glycols, among others, are used (there has been no book to-date describing actual applications). Practical techniques and technologies for designing and manufacturing cosmetics are given, as well as theoretical knowledge, in each of Chaps. 3–5. Chapter 3 describes emulsification, which is particularly difficult. Chapter 4 is on sensory evaluations of cosmetic ingredients that determine consumer preference. Chapter 5 describes how to mix and compose formulas for each kind of commercial cosmetic product, among which are featured soap-based cleansers, shampoos, conditioners, creams, hair-care products, hair dyes, agents for permanent waves, and curling lotions.

In summary, our work endeavors to instruct and inspire all enterprising cosmetics technicians to pursue further their passion for the industry and apply themselves in the following three areas: (1) the development and manufacture of hair-care and skin-care cosmetics, (2) the evaluation of raw materials, and (3) the design of cosmetics and consulting.

The potential for innovation is great in Japan's cosmetics industry. It is our hope that the high level of dedicated research continues and proliferates, especially in the fields of evaluation and development of high-grade, premium cosmetics products.

Tokyo, Japan Kunio Shimada

Contents

Part I
Ingredients and Productions of Cosmetics

Chapter 1
Developing the Formulations of Cosmetics

Abstract Cosmetics, which are daily applied on a part or all parts of the human body, require special knowledge and care for designing formulations. Ingredients should be combined so as to achieve the aimed efficacies and effects and be appropriate for the purpose, body part on which the product is to be applied, and method of use. This chapter also exemplifies the ingredients used for toilet soaps, cleansers, shampoos, rinse-off hair conditioners, leave-on hair care products, creams, oily cosmetics that do not contain water, and facial toner. All cosmetic products need to be temperature resistant and stable over a long period of time. Use of preservatives, pH regulators, chelating agents, and antioxidants should be carefully investigated. Stability must also be checked for changes in color, smell, and viscosity. Antiseptic effects must last long and secondary pollution after opening the pack also needs to be prevented.

Keywords Efficacies and effects • Ingredients • Limitations on inclusion of ingredients • Prohibited ingredients

1.1 Classification and Categories of Cosmetics

Cosmetics are daily applied on a part or all parts of the human body. There are diverse products, whose properties and forms are designed based on the efficacy and effects, the body part on which they are applied, and the purpose and method of use. According to the regulations on the manufacturing and sales of cosmetics in Japan, cosmetics are classified into the categories as shown in Table 1.1. This chapter describes the basic concepts and knowledge of designing cosmetics formulations.

H. Iwata and K. Shimada, *Formulas, Ingredients and Production of Cosmetics: Technology of Skin- and Hair-Care Products in Japan*, DOI 10.1007/978-4-431-54061-8_1,

Table 1.1 Categories of cosmetics

Hair care products
Hair liquids, hair tonics, scalp-care products, hair dyes, shampoos, and hair conditioners
Skincare products
Toilet water, beauty essences, creams, lotions, suntan lotions, sunscreens, cleansers[a]
Shaving lotions (facial), shaving lotions (body), facial conditioners, facial packs
Cosmetic oils[b], body conditioners, massage lotions
Makeup products
Foundations, bases, facial powders, lipsticks, eye makeup products, rouges
Body makeup products
Perfumes, colognes
Perfumes, colognes
Others
Bath preparations, nail cosmetics, body powders

[a] Of cleansers, facial cleansers are those for mainly cleansing the face

[b] Cosmetic oils are those applied on the skin and do not include oils used on hair, such as camellia oil

1.1.1 Parts of the Body

Cosmetics are applied on the skin, hair, scalp, nails, lips, teeth, and eyelashes including the oral cavity and around the eyes. Recent trends of cosmetics development focus on products to be applied on body parts on which cosmetics were rarely used and those that have new values. Cosmetics developments in new fields require thorough understanding and investigation of the body part on which the cosmetics is to be applied, the materials used and formulations.

1.1.2 Efficacies and Effects

The Pharmaceutical Affairs Law prescribes the range of cosmetics efficacy as shown in Table 1.2 (Pharmaceutical and Medical Safety Director Notification No. 1339, December 28, 2000).

No other efficacy or effect can be advertised for cosmetics. Cosmetics must not "cure" or "be effective against" symptoms and must be mild. Efficacy expressions permitted to cosmetics include "keeps something," "prevents something," "gives something," and "repairs something."

1.1.3 Purpose of Use

Users use cosmetics to "cleanse," "beautify," "change the appearance," and "protect" the bodies and "make them attractive." It should be noted that the purposes of use may differ from the efficacy expressions under the Pharmaceutical Affairs Law.

Table 1.2 Range of cosmetics efficacy

1	Cleans the scalp and hair	29	Softens the skin
2	Prevents odor of the scalp and hair by giving fragrance	30	Firms up the skin
3	Keeps the scalp and hair healthy	31	Gives luster to the skin
4	Makes the hair springy	32	Smoothens the skin
5	Moisturizes the scalp and hair	33	Facilitates shaving
6	Keeps the scalp and hair moist	34	Cares the skin after shaving
7	Makes the hair soft and smooth	35	Prevents prickly heat (by applying powder)
8	Makes the hair easy-to-comb	36	Prevents suntan
9	Keeps the hair glossy	37	Prevents blotches and freckles by sunburn
10	Makes the hair glossy	38	Gives fragrance
11	Removes dandruff and itchiness	39	Protects nails
12	Suppresses dandruff and itchiness	40	Keeps nails healthy
13	Supplements and keeps moisture and oil in the hair	41	Moisturizes nails
14	Prevents trichorrhexis and split ends of hair	42	Prevents dry lips
15	Helps arrange hair and keeps the hair style	43	Smoothens the texture of the lips
16	Prevents static charge of the hair	44	Moisturizes the lips
17	Cleanses the skin (by removing the impurities)	45	Keeps the lips healthy
18	Prevents acne and prickly heat (by cleansing the skin) (cleanser)	46	Protects the lips. Prevents the lips from drying
19	Prepares the skin	47	Prevents chapped lips by keeping the lips moist
20	Smoothens the skin	48	Smoothens the lips
21	Keeps the skin healthy	49	Prevents dental caries (toothpastes used for brushing the teeth)
22	Prevents dry skin	50	Whitens the teeth (toothpastes used for brushing the teeth)
23	Contracts the pores of the skin	51	Removes dental plaque (toothpastes used for brushing the teeth)
24	Moisturizes the skin	52	Cleans the oral cavity (toothpastes)
25	Supplements and keeps moisture and oil in the hair	53	Prevents mouth odor (toothpastes)
26	Keeps the skin flexible	54	Removes tobacco stains from teeth (Toothpastes used for brushing the teeth)
27	Protects the skin	55	Prevents tartar from accumulating on teeth (toothpastes used for brushing the teeth)
28	Prevents the skin from drying	56	Obscures wrinkles caused by drying of the skin

Permitted efficacy expressions of cosmetics, which do not need approval by the Pharmaceutical Affairs Law
(Pharmaceutical and Medical Safety Director Notification No. 1339, December 28, 2000)

1.1.4 Method of Use

Cosmetics are applied on a part or all parts of the body and are either left on, wiped off, or rinsed off. Most cosmetics are either rinsed off (rinse-off cosmetics) or left on (leave-on cosmetics); and the difference determines the safety level required to the products and the ingredients used. It should also be noted that there are restrictions on components and the concentrations depending on the part of the body on which the cosmetics is to be applied. See the Appendices of the Japanese Standards for Cosmetics of Japan at the end of this chapter.

1.1.5 Properties and Product Forms

The properties and forms of cosmetics are decided based on the body part on which the product is to be used and whether is it to be rinsed off, wiped off, or left on, and so that users feel easy to use. Marketing aspects, such as containers and proposed lifestyles, may also be reflected in the properties and product forms.

Main product forms include liquids, viscous liquids, milky lotions, creams, gels, waxes, solids, powders, and aerosols.

1.2 Constituents of Cosmetics

Constituents of cosmetics include water, oils, silicones, surfactants, polymers, polyhydric alcohols, saccharides, organic solvents, acid and alkali salts, inorganic and organic powders, pigment colors, amino acids, proteins, plant extracts, vitamins, ultraviolet absorbers, chelating agents, preservatives, antioxidants, oxidizing and reducing agents, and aromatic essential oils.

It is convenient to classify the constituents into those that give the product form, stabilize the product, have efficacy and effects, and act on the senses of users. Ingredients of cosmetics are combined so as to achieve the aimed efficacies and effects and be appropriate for the purpose, body part on which the product is to be applied, and method of use. The ingredients are outlined below and described in detail in Chap. 2.

1.2.1 Ingredients for Giving the Product Form

The constituents that give a product its form are water, oils, surfactants, silicone, polyhydric alcohols, polymers, and powders. They constitute the major part of a cosmetic product, determine its form, and affect the feel of use (Table 1.3).

Table 1.3 Ingredients for giving product forms

Principal materials	Principal objective of use
Oils	
Hydrocarbons, waxes, esters, vegetable oils and fats, higher alcohols, fatty acids	Oily ingredient, emulsifying aid
Surfactants	
Anionic surfactants, amphoteric surfactants, cationic surfactants, nonionic surfactants	Detergent, foaming agent, emulsifier, solubilizer, penetrating agent, softening agent, antistatic additives, antimicrobial agent
Silicones	Slip agent, antifoaming, texture improvement
Polyhydric alcohols	Moisturizing agent, stabilizer, solvent, solubilizing aid
Polymers	Thickener, setting agent, emulsion stabilizer, texture improvement
Inorganic powders	Powder, covering, coloring

1.2.2 Ingredients for Stabilizing the Products

Cosmetics contain ingredients for stabilizing the product and preventing deterioration. They are essential because cosmetics are used over a long period of time after opening the packet and have risks of degeneration and deterioration. Because cosmetics are applied on delicate parts of the body, such as the skin, they are required to be highly stable. The stabilizers are classified into antimicrobial agents, pH control chemicals, antioxidants, and chelating agents.

1.2.3 Ingredients for Giving Efficacies, Effects, and Concepts

Cosmetics are not allowed to be advertised for efficacy or effect, but ingredients that give efficacy and effects can be added. Such ingredients are also essential for presenting the concepts and purposes of the product. Care should be taken that expressions are strictly within the permitted "Range of cosmetics efficacy."

Each cosmetics manufacturer uses original combinations of ingredients that give efficacy, effects, and concepts. Manufacturers also use original components and advertise the use to attract users. The ingredients can be broadly classified into plant extracts and herbal medicine components, microbial-derived ingredients, proteins and amino acids, ceramides, and vitamins.

1.2.4 Ingredients Acting on the Senses of Users

Besides the major ingredients listed above, most cosmetics contain coloring and scenting agents to act on the senses of users. Coloring agents are used in hair dyes and also to give colors to cosmetics products. Only the latter is described in this book.

1.3 Categories of Cosmetics and Product Forms

The ingredients of cosmetics are combined so as to give the aimed product form, which depend on the category of cosmetics (shampoos, conditioners, creams, facial toner, body shampoo, lipsticks, foundations, etc.), and to match the purpose of the product. The ingredients used are exemplified below separately for toilet soaps, cleansers, shampoos, rinse-off hair conditioners, leave-on hair care products, creams, oily cosmetics that do not contain water, and facial toner.

1.3.1 Toilet Soaps and Cleansers

Toilet soaps and cleansers are for removing filth from the skin. There are toilet soaps, which are solid, and body shampoos and facial cleansers, which are either cream or liquid. Fatty acid salts are the most widely used component. Some products contain anionic or amphoteric surfactants as the cleansing component, in most cases together with fatty acid salts. Glycols such as glycerin are used as moisturizer. Higher alcohols and nonionic surfactants are combined to make the product creamy. Polymers are used to increase viscosity (Table 1.4).

1.3.2 Shampoos

Shampoos are for cleansing the scalp and hair. Anionic surfactants are used as the major cleansing agent, and amphoteric surfactants are also combined as the second cleansing agent. Nonionic surfactants such as alkanol amides stabilize foams and increase viscosity. Cationic polymers increase viscosity and improve the feel of use. These four main constituents determine the product form and the feel of use. Silicones are also widely used to improve the touch (Table 1.5).

1.3.3 Conditioners

Conditioners are applied after shampooing hair and are also called "hair treatment," "hair pack," and "hair rinse" depending on the contents of proteins, amino acids, and other effective components for repairing hair damage. The main purpose of conditioners is to improve the touch of hair and repair damage. Cationic surfactant of quaternary ammonium salt is always added to emulsify the product and improve the touch. Higher alcohols and all kinds and natural oils and fats can also be added to make the product creamy and improve the feel of use. Silicone is indispensable for giving a smooth touch to the hair. The most common form of conditioners is creams of various degrees of viscosity (Table 1.6).

Table 1.4 Basic formulations of soaps and cleansers

Toilet soap		
Fatty acids or oils	60–70%	Beef tallow: 80%, coconut oil: 20%
Glycerin	0–8%	Also add sucrose or other saccharides as necessary
Sodium hydroxide	18–22% of the fatty acids or oils	
Chelating agent (EDTA-4Na)		(EDTA-4Na)
Liquid soap		
Fatty acids or oils mainly consisting of lauric acid	5–25%	Emulsifier, touch improver
Glycerin	1–6%	Stabilizer, viscosity adjuster, moisturizer
Chelating agent (EDTA-4Na)		(EDTA-4Na, sodium citrate)
Potassium hydroxide	20–23% of the fatty acids or oils	
Anionic surfactants		Foaming agent
Amphoteric surfactants		Stimulus reducer
Nonionic alkanol amide surfactants		Thickener, foam stabilizer
Preservative		
Cream cleanser		
Fatty acids	5–20%	Detergent
Glycerin	1–6%	Moisturizer, stabilizer
Sodium hydroxide	20–23% of the fatty acids or oils	
Anionic surfactants	0–20%	Detergent, foaming agent
Amphoteric surfactants	0–12%	Touch improvement, hair protection, and stimulus reduction
Nonionic surfactants	3–6%	Thickener, foam stabilizer
Chelating agent		
Preservative		
Purified water		

Table 1.5 Basic formulations of shampoos (and body soaps)

Fatty acids	0–25%	Detergent
Anionic surfactants	3–9%	Detergent, foaming agent
Amphoteric surfactants	1–4%	Detergent, stimulus reducer, foaming agent
Nonionic surfactants	1–6%	Thickener, foam stabilizer
Glycol	1–6%	Stabilizer, viscosity adjuster, moisturizer
Polymers	0–1%	Thickener, touch improver
Neutralizer, pH adjuster		
Chelating agent		
Preservative		
Water		

Table 1.6 Basic formulations of conditioners

Cationic surfactants	1–5%	Hair softener, emulsifier, touch improvement
Higher alcohols	2–10%	Thickener, cream base
Liquid oils	0–10%	Oil, touch improvement
Nonionic surfactants	0–1%	Emulsification aid
Silicone	1–10%	Making the hair silky
Polymers	0.5% or less	Stabilizer, touch improver
Preservative		
pH adjuster		

1.3.4 Creams

Creams are for caring the skin of the face, hands, and other parts of the body. Product forms may be creams, milky lotions and gels. The purposes are diverse and include repairing the skin from damage, moisturizing the skin, whitening the skin and blotches, and protecting the skin from ultraviolet rays. The oily components, which give the efficacy and determine the feel of use, may be carbohydrates, esters, and higher alcohols. Nonionic surfactant and fatty acids are mainly used to emulsify the oily components. Polymers are added to stabilize the product form. Polyhydric alcohols are for moisturizing, and alkalis are combined to neutralize fatty acids (Table 1.7).

1.3.5 Leave-on Hair Care Products

There are hair care products of diverse purposes, such as for caring hair, improving the texture and styling. Product forms are also diverse and range from liquids to solids. There are hair mists, hair creams, waxes, styling agents, and hair oils. Constituents vary depending on purpose and product form and can be all kinds of surfactants, oil, polymer, and glycol.

Hair mists mainly consist of cationic surfactants and glycols. Hair creams contain oils, silicone, fatty acids (as emulsifier), and either anionic or cationic surfactants. Hair mists and creams also contain nonionic surfactants for emulsification. Hair waxes contain wax, micro crystalline waxes, fatty acids, nonionic surfactants, and glycols. Polymers are added to achieve hair setting and styling performances. Hair styling agents are gels or liquids that contain polymers. Hair oils are either silicones plus a small amount of plant oils or plant oils plus hydrocarbons (Table 1.8).

1.3.6 Cosmetics Mainly Consisting of Oil

Cosmetic products that mainly consist of oils include cleansing oils, self-emulsifying bath oils, body massage oils, hair oils, lipsticks, and solid lip rouges. Main constituents

Table 1.7 Basic formulations of creams

Cream		
Fatty acids	0–8%	Emulsifier, thickener
Higher alcohols	0–4%	Thickener, base of the cream
Esters	0–10%	Oil, touch adjuster
Carbohydrates	0–10%	Oil
Nonionic surfactant	1–4%	Emulsifier
Polymers	0.5% or less	Stabilizer
Preservative		
Alkali agent		
Water		
Gel		
Glycol	2–10%	Moisturizer
Ethanol	0–5%	Solvent
Polymers	0.3–2%	Thickener, gelling agent
Neutralizer	0.1–1%	Alkali
Preservative		
Water		
Gel		
Surfactants	20–50%	Thickener, gelling agent
Glycol	2–10%	Solubilizer, touch improver
Higher alcohol	0–10%	Thickener
Preservative		
Water		

are oils of any kind and nonionic surfactants. The water content is very small if there is any. Formulation design involves combining oils that differ in polarity and melting points and adding nonionic surfactant HLB (Table 1.9).

1.3.7 Facial Toner

Facial toner is designed by combining moisturizing agents so as to moisturize, soften, smoothen, firm up and give luster to the skin. It may also contain vitamin C derivatives for whitening, and fermentation extracts are included in some quasi-drug products for antiaging and wrinkle prevention. Many products also contain ceramides, vitamins, sterols, vegetable extracts, and other efficacious components.

Basic constituents are water, glycerin, and moisturizing agents, such as 1,3-butyl glycol, hydrolyzed collagen, amino acids, PCA-Na, and saccharides. Product forms may be transparent liquid, transparent and viscous liquid, and milky lotion.

Table 1.8 Basic formulations of leave-on hair care products

Hair cream (soap emulsion type)		
Stearic acid	3–8%	
Higher alcohols	3–8%	Thickener, good touch
Esters	0–3%	Touch adjuster
Nonionic surfactant	0–1%	Emulsification aid
Silicone	1–8%	To make the hair silky
Anionic polymer	0–0.5%	Stabilizer
Preservative		
Alkali agent		
Hair treatment cream		
Cationic surfactants	1–5%	Emulsifier, hair conditioner and smoother
Higher alcohols	3–6%	Thickener, good touch
Esters	0–3%	Touch adjuster
Nonionic surfactants	0–1%	Emulsification aid
Silicone	1–8%	To make the hair silky
Vegetable oils and fats	0–1%	To make the hair moist and soft
Cationic polymers	0–1.2%	Touch improver, stabilizer
Antioxidant	0–0.1%	
Preservative		
pH adjuster		
Hair mist		
Glycol	2–6%	Touch improver, moisturizer, stabilizer
Cationic surfactants	0–1%	To make the hair soft and smooth
Esters	0–5%	
Nonionic surfactant	0–2%	Emulsifier
Setting polymers	0–15%	Styling and setting agent
Ethanol	5–20%	Stabilizer, solubilizer
pH adjuster		
Preservative		
Hair wax		
Fatty acids	2–8%	Stearic acid, hydroxystearic acid
Waxes	2–8%	Candelilla wax, micro crystalline wax
Esters	0–9%	
Higher alcohol	0–4%	
Nonionic surfactants	2–6%	Emulsifier
Glycol	0–6%	
Silicone	0–5%	
Preservative		
pH adjuster		
Hair gel		
Glycol	2–6%	
Ethanol	0–10%	
Thickening polymers	0.3–1.5%	Mainly carboxy vinyl polymer
Alkali agent	0.1–1%	
Preservative		

Table 1.9 Basic formulations of cosmetics mainly consisting of oil

Solid type		
Liquid oils	20–100%	
Waxes	0–30%	
Nonionic surfactants	0–25%	Emulsifier
Hair oil		
Silicone	90–100%	
Liquid oils	Minute to small amount	

Table 1.10 Points to note about cosmetic ingredients

Not irritate or be toxic to the skin
Not disturb the physiological functions of the skin
Be stable and not undergo changes in color and smell
Not separate, not coagulate, not precipitate, and not undergo changes in property
Not undergo big changes in viscosity by passage of time or temperature changes
Suppress the development of bacteria and secondary pollution

1.4 Points to Note for Using Ingredients in Cosmetics

The Japanese Standards for Cosmetics (enacted in April 2001) prescribes the prohibited ingredients for cosmetics (Appendix 1) and limitations on inclusion of ingredients (Appendix 2), preservatives (Appendix 3), and UV absorbers (Appendix 4) to ensure the quality and safety of cosmetics. For each ingredient, there is a limitation on the amount that can be included depending on the body part the product is to be used and the method of use (to be rinsed off or left on). There are also restrictions on materials of biological origin. The former Japanese Standards for Cosmetic Ingredients should also be referred to because there were ingredients whose inclusion was restricted due to safety problem.

All cosmetic products need to be temperature resistant and stable over a long period of time. Use of preservatives, pH regulators, chelating agents, and antioxidants needs to be investigated. Stability must be checked for changes in color, smell, and viscosity. Antiseptic effects must last long, and secondary pollution after opening the pack also needs to be prevented. Other points to note on the ingredients of cosmetics are summarized in Table 1.10.

Standards for Cosmetics

Appendix 1

Ingredients Prohibited in All Types of Cosmetics

1. 6-Acetoxy-2,4-dimethyl-m-dioxane
2. Antihistamines except those of aminoether type (such as diphenhydramine)
3. Ingredients derived from cattle and other ruminants raised in an area other than the following countries: USA, Argentina, India, Uruguay, El Salvador, Australia, Canada, Kenya, Costa Rica, Colombia, Singapore, Swaziland, Chile, Nigeria, Namibia, Nicaragua, New Zealand, Pakistan, Panama, Paraguay, Brazil, Botswana, Mauritius
4. Ingredients produced from the following parts of cattle or other ruminants: Pituitary, thymus, dura mater, pineal gland, spinal cord, placenta, intestine, brain, cerebrospinal fluid, spleen, adrenal glands, tonsils, eyes, lymph nodes
5. Hormones and those derivatives except estradiol, estrone, and ethinylestradiol
6. Vinyl chloride monomer
7. Methylene chloride
8. Bismuth compounds other than bismuth oxychloride
9. Hydrogen peroxide
10. Cadmium compounds
11. Sodium perborate
12. Chloroform
13. Progrenolone acetate
14. Dichlorophene
15. Mercury and its compounds
16. Strontium compounds
17. Sulfamide and its derivatives
18. Selenium compounds
19. Nitrofuran-type compounds
20. Hydroquinone monobenzylether
21. Halogenated salicylanilide
22. Vitamin L1 and Vitamin L2
23. Bithionol
24. Pilocalpine
25. Pyrogallol
26. Inorganic fluorine compounds
27. Pregnanediol
28. Local anesthetics such as procaine
29. Hexachlorophen
30. Boric acid
31. Formalin
32. Methyl alcohol

Appendix 2

Ingredients Restricted in All Types of Cosmetics

Ingredient name	Maximum amount of ingredient per 100 g
Aluminum chlorhydroxy allantoinate	1.0 g
Cantharides tincture, ginger tincture, or capsicum tincture	1.0 g as total
Phenyl salicylate	1.0 g
Polyoxyethylene laurylether (8–10 E.O.)	2.0 g

Ingredients Restricted According to Types or Intended Purposes of Cosmetics

Ingredient name	Maximum amount of ingredient per 100 g
Aerosol agents	
Zirconium	Prohibited
Cosmetics to be washed away immediately after use such as soap or shampoo	
Thiram	0.50 g
Cosmetics other than those washed away immediately after use such as soap or shampoo	
Undecylenic acid monoethanolamide	Prohibited
Thiram	0.30 g
Zinc *p*-phenolsulfonate	2.0 g
2-(2-Hydroxy-5-methylphenyl) benzotriazole	7.0 g
Sodium lauroyl sarcosinate	Prohibited
Cosmetics used in cephalic, mucosa part or the oral cavity, and cosmetics used in other parts, containing lower aliphatic monoalcohols (exclude those containing the said alcohols added to dissolve ingredients in the said cosmetics)	
Estradiol, estrone, and ethinylestradiol	20,000 IU as total
Cosmetics other than those used in cephalic, mucosa part or the oral cavity, containing no lower aliphatic mono-alcohols (include those containing the said alcohols added to dissolve ingredients in the said cosmetics)	
Estradiol, estrone, and ethinylestradiol	50,000 IU as total
Cosmetics used in only cephalic part	
Aminoether type antihistamines	0.010 g
Cosmetics other than those used only in cephalic part	
Aminoether-type antihistamines	Prohibited
Toothpaste	
Sodium lauroyl sarcosinate	0.50 g
Compounds to be used with the purpose of emulsifying beewax or white beewax	
Sodium pyroborate	0.76 g (limited not to greater than 1/2 amount of beewax and white beewax)
Compounds to be used for purposes other than emulsifying beewax or white beewax	
Sodium pyroborate	Prohibited

Ingredients Restricted According to Types of Cosmetics

	Maximum amount (g) of ingredient per 100 g	
	Cosmetics not used for mucosa	Cosmetics that may be used for mucosa
Ubidecarenone	0.03	Prohibited

Appendix 3

Ingredients Restricted in All Types of Cosmetics

Ingredient name	Maximum amount (g) of ingredient per 100 g
Benzoic acid	0.2
Benzoate	1.0 as total
Alkyldiaminoethylglycine hydrochloride	0.20
Photosensitizing dyes	0.0020 as total
Chlorcresol	0.50
Chlorobutanol	0.10
Salicylic acid	0.20
Salicylate	1.0 as total
Sorbic acid and sorbate	0.50 as total
Dehydroacetic acid and dehydroacetate	0.50 as total
Trichlorohydroxydiphenylether (Triclosan)	0.10
p-Oxybenzoic acid esters and their sodium salts	1.0 as total
Phenoxyethanol	1.0
Phenol	0.10
Sodium lauryldiaminoethylglycine	0.030
Resorcin	0.10

Ingredients Restricted According to Types of cosmetics[a]

	Maximum amount (g) of ingredient per 100 g		
Ingredient name	Cosmetics not used for mucosa and to be washed away	Cosmetics not used for mucosa and not to be washed away	Cosmetics that may be used for mucosa
Zinc, ammonia, and silver-substituted zeolite[b]	1.0	1.0	
Pantothenyl ethylether benzoate	○	0.30	0.30

(continued)

Ingredients Restricted According to Types of cosmetics[a] (continued)

	Maximum amount (g) of ingredient per 100 g		
Ingredient name	Cosmetics not used for mucosa and to be washed away	Cosmetics not used for mucosa and not to be washed away	Cosmetics that may be used for mucosa
Isopropylmethylphenol	○	0.10	0.10
Cetylpyridinium chloride	5.0	1.0	0.010
Benzalkonium chloride	○	0.050	0.050
Benzethonium chloride	0.50	0.20	
Chlorhexidine hydrochloride	0.10	0.10	0.0010
o-Phenyl phenol	○	0.30	0.30
Sodium *o*-phenylphenate	0.15	0.15	
Chlorhexidine gluconate	○	0.050	0.050
Cresol	0.010	0.010	
Chloramine T	0.30	0.10	
Chlorxylenol	0.30	0.20	0.20
Chlorphenesin	0.30	0.30	
Chlorhexidine	0.10	0.050	0.050
1,3-Dimethylol-5, 5-dimethylhydantoin	0.30		
Alkylisoquinolinium bromide	○	0.050	0.050
Thianthol	0.80	0.80	
Thymol	0.050	0.050	○[b]
Trichlorocarbanilide	○	0.30	0.30
p-Chlorphenol	0.25	0.25	
Halocarban	○	0.30	0.30
Hinokitiol	○	0.10	0.050
Zinc pyrithione	0.10	0.010	0.010
Iodopropynyl butylcarbamate	0.02	0.02	0.02
Methylchloro isothiazolinone and methyl isothiazolinone solution[d]	0.10		
N,N″-Methylenebis[*N′*-(3-hydroxymethyl-2,5-dioxo-4-imidazolidinyl)urea]	0.30		
p-Dimethylaminostyryl heptyl methyl thiazolium iodide	0.0015	0.0015	
Methyl isothiazolinone	0.01	0.01	
Silver-copper zeolite[e]	0.5	0.5	
Polyaminopropyl biguanide	0.1	0.1	0.1

[a] Blank indicates that it is prohibited to be used, and ○ indicates that there is no upper limit for the amount of ingredient

[b] It indicates the compound containing 0.2–4.0% as silver and 5.0–15.0% as zinc when it is exposed to strong heat

[c] It can be contained in cosmetics used for mucosa and only for oral cavity

[d] It indicates the aqueous solution containing 1.0–1.3% of 5-chloro-2-methyl-4- isothiazolin-3-one and 0.30–0.42% of 2-methyl-4-isothiazolin-3-one

[e] It indicates the compound containing 2.7–3.7% as silver and 4.9–6.3% as copper when it is exposed to strong heat

Appendix 4

Ingredients Restricted in All Types of Cosmetics

Ingredient name	Maximum amount (g) of ingredient per 100 g
p-Aminobenzoic acid and its esters	4.0 as total

Ingredients Restricted According to Types of Cosmetics[f]

	Maximum amount (g) of ingredient per 100 g		
Ingredient name	Cosmetics not used for mucosa and to be washed away	Cosmetics not used for mucosa and not to be washed away	Cosmetics that may be used for mucosa
Octyl salicylate	10	10	5.0
Homomenthyl salicylate	10	10	10
Methyl-2, 5-diisopropylcinnamate	10	10	
Cinoxate	○	5.0	5.0
Glyceryl mono-2-ethylhexonoate di-*p*-methoxycinnamate	10	10	
Dihydroxydimethoxybenzophenone	10	10	
Sodium dihydroxydimethoxybenzophenone disulfonate	10	10	
Dihydroxybenzophenone	10	10	
1-(3,4-dimethoxyphenyl)-4,4-di-methyl-1,3-pentanedione	7.0	7.0	
Dimethoxybenzylidenedioxoimidazolidine 2-ethylhexyl propionate	3.0	3.0	
Tetrahydroxybenzophenone	10	10	0.050
2,4,6-Tris [4-(2-ethylhexyloxycar bonyl) anilino]-1,3,5-triazine	5.0	5.0	
Methylbis(trimethylsiloxy)silyl iso-pentyl trimethoxycinnamonate	7.5	7.5	2.5
Amyl *p*-dimethylaminobenzoate	10	10	
2-Ethylhexyl *p*-dimethylaminobenzoate	10	10	7.0
Isopropyl *p*-methoxycinnamate and diisopropyl cinnamate ester mixture[g]	10	10	
2-Ethylhexyl *p*-methoxycinnamate	20	20	8.0
2-Hydroxy-4-methoxybenzophenone		5.0	5.0
Hydroxymethoxybenzophenone sulfonate and its trihydrate	10[h]	10[h]	0.10[h]
Sodium hydroxymethoxybenzophenone sulfonate	10	10	10

(continued)

Ingredients Restricted According to Types of Cosmeticsf (continued)

Ingredient name	Maximum amount (g) of ingredient per 100 g Cosmetics not used for mucosa and to be washed away	Cosmetics not used for mucosa and not to be washed away	Cosmetics that may be used for mucosa
Phenylbenzimidazole sulfonic acid	3.0	3.0	
4-tert-butyl-4′-methoxydibenzoylmethane	10	10	10
4- (2-β-glucopyranosiloxy) propoxy-2-hydroxybenzophenone	5.0	5.0	
Terephthalylidene dicamphor sulfonic acid	10	10	
Ferulic acid	10	10	
2-Cyano-3,3-diphenyl-2-propenoic aid 2-ethylhexyl ester (Octocrylene)	10	10	10
Drometrizole trisiloxane	15	15	
2,2′-methylenebis(6-(2 H-benzotriazole-2-yl)-4-(1,1,3,3-tetramethylbutyl) phenol	10	10	
2-[4-(diethylamino)-2-hydroxybenzyl] benzoic acid hexylester	10	10	
Dimethicodiethyl-benzal malonate	10	10	10

[f] Blank indicates that it is prohibited to be used, and ○ indicates that there is no upper limit for the amount of ingredient

[g] It indicates the compound containing 72.0–79.0% of isopropyl *p*-methoxycinnamate, 15.0–21.0% of ethyl 2,4-diisopropyl cinnamate, and 3.0–9.0% of methyl 2,4-diisopropyl cinnamate

[h] It is calculated as the total amount of hydroxymethoxybenzophenone sulfonate

Chapter 2
Raw Materials of Cosmetics

Abstract Raw materials used in cosmetics are described in this chapter.

Oils used in cosmetics are hydrocarbons, animal and vegetable oils and fats, waxes, esters, higher alcohols, fatty acids, and silicones. The chemical structures and physical properties of oily ingredients and their effects on the feel of use and texture of products are described.

Surfactants have both hydrophilic and lipophilic groups in a single molecule. The properties of each kind of surfactant are outlined, and the characteristics of major surfactants used in cosmetics are described.

Polymers are widely used in cosmetics mainly for thickening and also for stabilizing, hair setting and improving the feel of use. The viscosity of polymers depends on the molecular weight, and the chemical structures of the hydrophilic group and alkyl chain determine the properties.

Glycols have polarities that are intermediate of surfactants and are determined by the OH group and the number of carbons. They can be used for various purposes by using the properties and are very useful in preparing cosmetics.

Keywords Animal and vegetable • Glycols • Oily ingredients • Polymers • Surfactants

2.1 Oils

Most oils used in cosmetics can be broadly classified by chemical structure into hydrocarbons, animal and vegetable oils and fats, waxes, esters, higher alcohols, fatty acids, and silicones.

The melting point, which is one of most important properties, is determined by the molecular weight; and oils of larger molecular weights melt at higher temperatures.

H. Iwata and K. Shimada, *Formulas, Ingredients and Production of Cosmetics: Technology of Skin- and Hair-Care Products in Japan*, DOI 10.1007/978-4-431-54061-8_2,

Polarity, which is another property, is determined by the chemical and molecular structures (presence of double bonds and side chains and the kind of functional group) and the number of oxygen atoms. These properties determine the compatibility with other ingredients and govern the viscosity, consistency, and feel of use of the products. Oily ingredients also affect the appearance, hardness, and spread of the cosmetics. Therefore, oily ingredients can be said to determine the feel of use and texture of cosmetics products. This chapter describes the chemical structures and physical properties of oily ingredients and their effects on the feel of use and texture of products.

2.1.1 Hydrocarbons

Hydrocarbons do not possess oxygen atoms, are saturated, and are thus nonpolar. The straight or branched chains consisting of carbon and hydrogen atoms can be combined in all cosmetics because they are scarcely oxidized and are hardly affected by pH changes and oxidation and reducing agents. They are useful as nonpolar oil for making emulsions. The chemical structure, melting point, and molecular weight of the hydrocarbon to be included should be examined for consistency with the purpose, properties, and feel of use of the product.

2.1.1.1 Isododecane

Isododecane is unsaturated 2,2,4,6,6-pentamethylheptane (mw:170.3), which is free of aromatics, is highly hydrorefined and is easy to dry. It is highly compatible with dimethicone and is a useful solvent for highly polymerized silicones. It is flammable and should be kept away from flames.

It gives a very light touch, and almost no feel is felt when applied on the skin.

2.1.1.2 Dodecane and Tetradecane

Dodecane and tetradecane are straight-chain C12 and C16 hydrocarbons of vegetable origin (boiling point: 177°C, ignition point: 71°C, specific gravity: 0.74–0.76, viscosity: 1–3[mPas]). They are flammable.

2.1.1.3 Squalane

Commercially available squalane is produced from deep sea sharks, is liquid, melts at a low temperature, and has a structure in which both ends of hexamethyl tetracosane $C_{30}H_{62}$ (mw: 422.8) branch. It gives a light touch and thus is suitable for leave-on cosmetics.

2.1.1.4 α-Olefin Oligomers

α-Olefin oligomers are straight chains of C4–12 and resemble squalane. Like squalane, they give a light texture and are suitable for leave-on cosmetics.

2.1.1.5 Liquid Paraffin

Liquid paraffin is straight-chain hydrocarbon of C16–32, consisting of paraffin and naphthene, and has a relatively low molecular weight. It gives a stronger oily and moisturized feeling than squalane and α-olefin oligomer. It is suitable for rinse-off cosmetics and for increasing the content of oily liquid hydrocarbon.

2.1.1.6 Vaseline (Petrolatum)

White vaseline is saturated hydrocarbon of C24–34 prepared by dissolving petroleum in solvent and collecting the precipitated crystals. The melting point is at 38–60°C, and it is paste-like at the room temperature. It is suitable as an oily ingredient for creams, bases, and solid cosmetics that do not contain water.

2.1.1.7 Isoparaffin

Isoparaffin is hydrogenated copolymers (C5–10) of iso-butene and n-butene. The transparent viscous liquid is classified by molecular weight into light liquid isoparaffin, light isoparaffin, liquid isoparaffin, and heavy liquid isoparaffin.

Light liquid isoparaffin has low viscosity, dries easily and gives a very light texture. Due to its high compatibility with dimethicone, it can be used as an organic solvent. Heavy liquid isoparaffin is viscous and gives a strong sticky, adsorbing and heavy feeling. However, when combined in an emulsion, the heavy feeling disappears, and the product gives a stronger silky feeling than liquid paraffin.

2.1.1.8 Microcrystalline Waxes

Microcrystalline waxes are isoparaffin and naphthenic hydrocarbons of C31–70 and have molecular weights of 450–1,000 and melting points of 60–85°C. Due to the branched chemical structure, the crystals are small.

Those that melt at 80°C give a non-adsorptive and rather silky feeling; and those that melt at higher temperatures give stronger persistent feeling and less silky texture. It is important to choose the melting point by the purpose of use. Waxes are used for giving the form to stick-type cosmetics such as lipsticks and are also widely used in hair waxes. Those of high melting points produce waxes that can fix the hair firm. They are also added into rinse-off hair conditioning products for enhancing the springiness.

2.1.1.9 Ceresin

Ceresin is normal paraffin of C29–35. The hard crystals are produced by purifying ozocerite. The melting point is 61–95°C. Despite the high melting point, it gives a light feeling (Table 2.1).

Table 2.1 Hydrocarbons used in cosmetics

Name	Basic structure, properties, composition
Isododecane	$C_{12}H_{26}$
Dodecane	Straight-chained hydrocarbon of C12
Tetradecane	Straight-chained hydrocarbon of C16
Squalane	$C_{30}H_{62}$
α-Olefin oligomer	Straight chain of C4–12
Liquid paraffin	C16–32
Isoparaffin	Light liquid isoparaffin, light isoparaffin, liquid isoparaffin, heavy liquid isoparaffin
Microcrystalline wax	Isoparaffinic and naphthenic hydrocarbons of C31–70
Isododecane	$C_{12}H_{26}$
Dodecane	C12
Copolymer of rapeseed and *Aleuritesfordii* oils	IV: 90, SV: 180
Isopentane	C_5H_{12}
α-Olefin oligomer	Straight chain of C4–12
Hydrogenated diterpenes	Hydrogenated copolymers iso-butenes and n-butenes. Light liquid isoparaffin of C5–10
	Light isoparaffin
	Liquid isoparaffin
Hydrogenated polyisobutene	Heavy liquid isoparaffin
Hydrogenated polydecene	Straight-chained olefin oligomer of C4–12. Degree of polymerization: 3–6
Squalane	$C_{30}H_{62}$. Hexamethyltetracosane. Derived from deep-water sharks
Ceresin	Produced by purifying ozokerite. Crystalline. Normal paraffin of C29–35. Melting point: 61–95°C
Paraffin	Normal paraffin of C16–40. Melting point 50–70°C
Polyisobutene	
Polyisoprene	Liquid polyisoprene
Polyethylene	Polymers of ethylene. Molecular weight: 2,000–5,000. Powders of polyethylene and high-melting-point polyethylene
Polystyrene	
Polybutene	Produced by decomposing and distilling naphtha and polymerizing butane and butene gases. Molecular weight: 500–5,500
Polypropylene	
Mineral oils (liquid paraffin)	Paraffin and naphthene. C16–32
Petrolatum	Precipitated crystals obtained by dissolving petroleum in solvent. C24–64. Paste-like substance
Vegetable squalane	$C_{30}H_{62}$. Produced by hydrogenating squalane extracted from vegetable oils such as olive, rice bran, wheat germ, and sesame oils
Isohexadecane	

2.1.2 *Vegetable Oils and Fats*

Animal and vegetable oils and fats are triglycerides, which are esters of three fatty acids and glycerin. They are highly polar due to the three ester bonds. Many kinds are available. They are included in large amounts as a chief ingredient of cosmetics, or may be included in a very small quantity as a conception ingredient.

To use them in cosmetics, the alkyl compositions should be understood, and appropriate oil should be selected based on purpose because the feel of use is determined by the alkyl composition and melting point. Vegetable oils and fats are especially prone to oxidization and changes in color by light, oxygen, and heat because the alkyl group has unsaturated fatty acids, which have a double bond(s) prone to oxidization. The iodine number is a useful index for knowing oxidization stability. Oils and fats that contain linoleic acid, which has three double bonds, have large iodine numbers and are especially prone to oxidization. Care should be taken in using natural fats and acids that have linoleic acid and large iodine numbers, and it is recommended to lower the pH or add antioxidants such as vitamin E and lecithin.

Oils and fats that contain lauric acid and/or myristic acid, which have short alkyl chains, give a light touch. Solid fats that contain large amounts of palmitic acid and/or stearic acid reduce lubrication. Liquid oils rich in unsaturated fatty acids, such as oleic acid, give a moisturized feeling. Table 2.2 lists representative vegetable and animal oils and fats.

It is easy to understand the properties of the vegetable and animal oils and fats, which helps designing cosmetics formulations, by classifying them by the length of the saturated alkyl chains, iodine number and the distributions (and contents) of unsaturated acids, such as oleic, linoleic and linolenic acids.

2.1.2.1 Vegetable Oils and Fats Rich in Linoleic and Linolenic Acids

Soybean oil and sunflower oil, which contain about 10% linolenic acid, have low melting points. Antioxidization measures should be taken when using them in cosmetics because the iodine number is large. Rapeseed oil has a relatively small iodine number because it contains more oleic acid that linolenic acid.

2.1.2.2 Vegetable Oils and Fats Rich in Oleic and Linoleic Acids

Safflower oil, which contains as much as 77% linoleic acid, and cotton seed oil and maize oil, which contain about 55% linoleic acid, require measures for preventing oxidization. Oils extracted from sesame and rice bran contain about 40% linoleic acid and 40% oleic acid and can be used in a large quantity in cosmetics by taking oxidization preventive measures.

2.1.2.3 Vegetable Oils and Fats Rich in Oleic Acid

Oils extracted from avocado, olive, camellia, and rice bran, and persic oil contain at least 70% oleic acid. They have relatively low iodine numbers of 80–110 and are stable against oxidation and are thus suitable for cleansing and massage oil products, to which a large amount of oil must be prescribed. Because they give a moderate moisturized feeling, they are also used in hair care cosmetics. Antioxidants such as vitamin E and its derivatives are essential for products mainly consisting of oils, and the long-term stability must be checked. The oils can be added in creams by lowering the pH to the lowest level. They give a moderate moisturized feeling.

2.1.2.4 Vegetable Oils and Fats Rich in Lauric Acid

Coconut and palm kernel oils contain approximately 50% lauric acid. They are prone to producing a unique sweet smell, and the stability should be confirmed before use. Solid oils give a light and dry feeling but are not widely used.

2.1.2.5 Vegetable and Animal Oils and Fats Mainly Consisting of Oleic and Palmitic Acids

Palm kernel oil, which contains about 50% palmitic acid and 40% oleic acid, melts at 45°C and is solid at room temperature. It has a small iodine number and is stable against oxidation. Because it is solid fat, it gives a relatively heavy touch and reduces lubrication. It is not suitable for cosmetics that need to be soft and smooth. However, it is useful for giving a light and dry feeling and is suitable for hand creams and other non-silky products. Animal fats of similar compositions are beef tallow and lard.

2.1.2.6 Vegetable Oils and Fats Containing Palmitoleic Acid

Macadamia nut oil is a rare vegetable oil that contains 25% palmitoleic acid. It has a small iodine number, a low melting point and relatively high oxidation stability. When used for hair care products, it provides a moisturized and soft feeling. It is an easy-to-use vegetable oil.

2.1.2.7 Vegetable Oils and Fats Containing Ricinoleic Acid

Castor oil contains approximately 90% ricinoleic acid and is highly viscous. Because of the heavy feeling, it is unfavorable for skin care products that need to be refreshing. Inclusion of several percent is suitable for leave-on hair treatment products that aim for giving a strong moisturized feeling.

2.1.2.8 Vegetable Oils and Fats Containing Oleic and Steric Acids

A typical fat of this group is shear butter. It contains 65% oleic acid and 20% stearic acid. It gives a moisturized and relatively heavy touch. Due to its small iodine number, it is stable against oxidization. It is useful for giving a moderate moisturized feeling to skincare and hair care products.

2.1.2.9 Vegetable Oils and Fats Containing Eicosenoic Acid

Meadowfoam seed oil consists of long-chain fatty acids of about 60% eicosenoic acid and 15% erucic acid. It is stable against oxidization and gives a moisturized and persistent feeling.

2.1.3 Waxes

Waxes are high-melting-point materials consisting of unsaponifiable compounds such as hydrocarbons, fatty acids, alcohols and esters of fatty acids and higher alcohols that possess at least 20 carbon atoms in the alkyl group. They include beeswax, carnauba wax, candelilla wax, rice wax, lanolin, and jojoba oil, among many others. Due to the high melting points, they are useful for producing hard creams, pastes or stick-shaped products, such as nonaqueous lipsticks. Representative waxes used in cosmetics are listed in Table 2.3.

Beeswax, which is collected from beehives, is the most widely used wax in cosmetics. It contains fatty acids and becomes creamy when neutralized by alkali. It is used for creams either alone or combined with stearic acid. It may also be used for concealers and protection creams, which are entirely made of oily ingredients, and also may be combined with petrolatum and lanolin. It gives a moisturized and waxy touch.

Carnauba wax has large numbers of carbons in both fatty acid and higher alcohol regions. It is used for solid cosmetics, particularly in products entirely made of oily ingredients, such as lipsticks and hair waxes. It is also used in some hair conditioner products to give a coated touch. It is difficult to emulsify and may require both cationic and nonionic surfactants.

Candelilla wax is used for solid cosmetics, lipsticks and hair waxes, which are entirely made of oily ingredients, and hair waxes to constitute 2–6%. The high melting point helps users to freely arrange the hair.

Rice wax is used for solid cosmetics, lipsticks and hair waxes, which are entirely made of oily ingredients, like candelilla wax.

Lanolin has been long and widely used in cosmetics. It contains hydroxy fatty acids as well as higher esters and has thus a water holding capacity. It is efficacious because it contains sterols and is suitable for skin creams to give a moisturized touch. It is also very suitable for hair care products because it gives a good and moisturized texture.

Jojoba oil is an ester consisting of unsaturated fatty acids and higher alcohol and is liquid at the room temperature. It is widely used in cosmetics products because it is stable against oxidization.

Table 2.2 Vegetable and animal oils and fats used in cosmetics [2]

INCI	Laurate	Myristate	Palmitate	Stearate	Oleate	Linoleate	Linolenate	Melting point	Refractive index	SV	IV	Characteristics
Persea gratissima (avocado) oil			6.9	0.6	77.3	10.8			1.74	192.6	94.4	Contains vitamins A and E and has high ultraviolet absorption capacity. Penetrates into and spreads particularly well on the skin
Prunus amygdalus dulcis (sweet almond) oil			6.7	1.2	66.3	22.3			1.46	188–200	92–105	Mainly consists of oleic acid and is slightly more unsaturated than olive oil. Low freezing point
Olea europaea (olive) fruit oil			9.8	3.2	73.8	11.1			1.456	185–197	75–90	Mostly consists of oleic acid and contains squalane
Theobroma cacao (cocoa) seed butter								35–36	1.457	185–199		Melts at around the body temperature (36–37°C). Has thus been used for lipsticks. Today, it is used for the base of creams
Seamum indicum (sesame) seed oil			8.8	5.3	39.2	45.8			1.475	185–195	103–118	
Oryza sativa (rice) bran oil			16.2	1.8	41.4	37.5	1.6		1.472	179–196	99–103	
Carthamus tinctorius (safflower) seed oil			6.8	2.5	12.6	77.4		–5	1.475	179–194	120–150	
Butyrospermum parkii (shea butter) fruit			3	20	65	10		28–45	1.458	160–190	50–80	Contains tocopherol and vitamin E and prevents oxidization before purification. Melts at around the body temperature. Is low viscous and lubricous and penetrates well. Contains 4–11% unsaponified matters

Glycine soya (soybean) oil			10.4	4	23.5	53.5	8.3	−8 to −7	1.472	188–196	114–138	Usually used like olive oil (but is less stable than olive oil). Applied on the skin to protect to it from irritants and drying
Zea mays (corn) germ oil			11.1	2.1	32.6	52.2	1.4	−8 to −10	1.474	187–198	88–147	Lowers the cholesterol content in the blood
Brassica campestris (rapeseed) oil, canola oil					57.9	21.8	11.3			183–197	87–107	Contains much oleic acid. Nondrying oil. Resembles olive oil. Contains 1.7% erucic acid
Prunus armeniaca (apricot) kernel oil, or *Prunus persica* (peach) kernel oil			7.2	2.3	74.8	15.6			1.471	188–199	91–110	Good-quality nondrying oil. Contains abundant oleic acid. Resembles almond oil
Elaeis guineensis (palm) kernel oil	5.3	14	6.8	2	16	Carpylic acid 3 Capric acid 4			21–24	1.45		Lauric oil. Contains more oleic acid than palm oil and has the advantages of both coconut and palm oils
Elaeis guineensis (palm) oil			49	2	41	7		30–50	1.466			Mainly consists of oleic and palmitic acids. Resembles beef tallow
Ricinus communis (castor) seed oil			1	3.1		4.4	Ricinoleic acid 89.6		1.478	176–187	81–91	Contains a large amount of ricinoleic acid and is thus very viscous. Attracts moisture. Widely used in hair cosmetics and soaps
Helianthus annuus (sunflower) seed oil			6.7	4		17.9	69.8	−18 to −16	1.47	186–194	113–146	Mainly consists of linoleic acid. High physiological activity
Grapeseed oil			11.1	3.3	21.2	61.4			1.473	180–196	107–143	
Macadamia ternifolia seed oil		1	7	3	55	2	Palmitoleic acid 25		1.47	190–200	70–80	Is a rare oil that contains at least 20% palmitoleic acid

(continued)

Table 2.2 (continued)

INCI	Laurate	Myristate	Palmitate	Stearate	Oleate	Linoleate	Linolenate	Melting point	Refractive index	SV	IV	Characteristics
Limnanthes alba (meadowfoam) seed oil	Icosanoic acid 2, eicosenoic acid 66, docosenoic acid 20, docosadiene acid 10									160–175	90–105	Consists of long-chain fatty acids of 60% eicosenoic acid and 15% erucic acid. Stable against oxidization
Gossypium (cotton) seed oil			20.1	2.4	18.9	56.5			1.474	189–197	88–121	Consists of 50% linoleic acid, 25% oleic acid and saturated fatty acids. Prone to oxidization due to high linolate content
Moringa pterygosperma seed oil												Agrees to the skin and gives smooth feeling. Further use in cosmetics expected. Superior oxidization stability
Cocos nucifera (coconut) oil	47	18	9.5	2.9	6.9	Carpylic acid 7.7 Capric acid 6.2		20–28	1.449	245–271	7–16	Relatively stable against oxidization but rather prone to hydrolysis. Used in creams, shampoos, and conditioners. Important raw materials for producing laurate and myristate. When used in soaps, dissolves well in cold water and lathers well. Easy to saponify by hydroxyl alkali solution
Arachis hypogaea (peanut) oil		11		2.9	42.2	34.7	2.6					Nondrying oil. Emollient effect. Equivalent penetration to olive and almond oil
Beef tallow		7	26.6	18.2	41.8	3.3			1.457	190–202	25–60	
Horse oil		3.2	24.9	5.5	35.5	10.8	9.5		1.463	195–204	71–86	Contains 7% palmitoleic acid
Mink oil		5	21	3	42	6			1.469	190–210	75–90	Contains 6% myristoleic acid and 18% palmitoleic acid

Table 2.3 Waxes for cosmetics [2]

	Base, characteristics, properties, composition	Specific gravity	Refractive index	Melting point (°C)	Acid value ratio[a]	Saponification value	Iodine number	Unsaponified matter content (%)
Orange raffia oil	Fish of the Trachichtyidae family living in the deep seas of South Africa and Australia. Contains 80–90% esters (oleil alcohol and oleic acid)	0.855(40)	–	–	0.37	98–108	73–89	48
Carnauba wax	Secretion from the leaves of carnauba palms in Brazil. Contains fatty acids of at least C24 and 85% of alcohol esters of C26, 28, and 30	0.990–1.001	1.469–72(40)	72–86	0.3–9.7	79–95	7–14	54–56
Candelilla wax	Stems of a plant inhabiting northwestern Mexico and Texas. Consists of 45% hydrocarbons, 29% esters, and 26% others	0.982–0.986	1.455–1.461	68–72		46–65	10–22	
Rice wax	Wax produced from rice bran. Ester of C22 and 24 fatty acids and C24 to 34 alcohols	–	–	70–83		70–160	20 or less	
Jojoba oil	Bush found in California and northwestern Mexico. Ester of unsaturated fatty acids and alcohols (C28–34: 83%)	0.853–0.875	1.455–1.475		0.75	80–110	70–100	50
Beeswax	Fatty acids (C16–18: 70%) and alcohols (C28–34: 83%)	0.961–0.964	1.456–1.458	61–66	5–7.6	80–101	4–12	55–58
Lanolin	Sheep wool. Mainly esters of hydroxyl fatty acids. Contains sterols	0.932–0.945	478–1.482	31–43	0.3–16	77–130	15–47	35–46

Reference: Courtesy of Nikko Chemicals

[a]Acid value ratio is the ratio of ester value to oxidization

2.1.4 Esters

Esters are oils consisting of fatty acids and alcohol, which are bound by ester bonds. Because they are combinations of fatty acids and alcohol, there are many ester compounds, which have one, two, or three ester bonds.

The properties vary depending on the structure of the alcohol and fatty acid regions, the number of carbons and the number of esters. In general, branch-chained esters give more moisturized and smooth feeling than straight-chained esters. Because there are two oxygen atoms at the center of the molecule, esters have intermediate polarity between those of hydrocarbons and higher alcohols, and therefore esters are compatible with other ingredients. Care should be taken against esterolysis, which occurs in alkaline conditions.

2.1.4.1 Esters of Straight-Chain Fatty Acid and Straight-Chain Lower Alcohol

This group includes isopropyl myristate, isopropyl palmitate, and butyl stearate. Because higher alcohols have low molecular weights, the properties and influences of the fatty acid component are dominant. Isopropyl myristate and isopropyl palmitate are particularly widely used.

2.1.4.2 Esters of Straight-Chain Fatty Acid and Straight-Chain Higher Alcohol

Cetylpalmitate and stearyl stearate are esters that have high melting points. When combined in creams, they give a dry, light and non-silky feeling rather than smooth and silky feeling. It can be combined with other oily ingredients of high melting points to provide a touch of firmness.

2.1.4.3 Esters of Branched-Chain Fatty Acid and Straight-Chain Alcohol

There are myristylethylhexanoate, cetylethylhexanoate, and cetylisostearate in this group. They are suitable for rinse-off and leave-on cosmetics, skincare and hair care products for giving smooth, silky, and moderate moisturized feelings. They are the most widely used esters.

2.1.4.4 Esters of Straight-Chain Fatty Acid and Branched-Chain Alcohol

This group includes ethylhexylmyristate, ethylhexylpalmitate, and isooctadecylpalmitate. They are used like the aforementioned esters consisting of branched-chain fatty acid and straight-chain alcohol, but provide a slightly strong feeling of oiliness. Because they have the characteristics of both straight-chain and branched-chain esters, they have balanced characteristics and are useful for binding oils of different polarities. They are widely used in cosmetics because they have low melting points, are liquid in general, and give smooth and silky feelings.

2.1.4.5 Esters of Branched-Chain Fatty Acid and Branched-Chain Alcohol

Isooctadecylisooctadecanoate and isononylisononate are highly compatible with silicone and form uniform liquid. They can also be used as a solvent for dimethicone. They give a light and silky feeling.

2.1.4.6 Esters of Fatty Acid and Polyhydric Alcohol

Glycerylmonostearate is a typical ester of this group. They are important as non-ionic surfactants and are widely used in cosmetics. They are effective emulsification aids. However, they deteriorate in alkali and cannot be formulated in high pH products.

2.1.4.7 Amphiphilic Esters of Carboxylic Acid and Alcohol

Each member of this group, such as cetyl lactate and triethylcitrate, has its own characteristics. They can be used as carriers of solvents and efficacious materials. Many amphipathic esters have been developed in recent years and used in diverse manners.

2.1.4.8 Dimer Acid Esters [1]

Dimer acid esters have four ester bonds surrounded by four alkyl groups and are relatively stable in acidic and alkaline conditions. They provide a strong sticky and viscous feeling. They are easy to use combined with another ester. When combined with emulsifiers or other oily ingredients, they produce a persistent but smooth and favorable feeling. Due to the chemical structures, they are stable and suitable for combining in alkaline agents [1].

2.1.4.9 Pentaerythritol Esters

Due to their complicated chemical structures, pentaerythritol esters give a moisturized feeling. There are various forms including monoesters of alcohol and organic acid or fatty acid, such as isopropyl palmitate and cetylethylhexanoate, and diesters [2].

2.1.5 *Higher Alcohols*

Higher alcohols have a straight, branched, or unsaturated chain of carbons with an OH group at an end. In general, higher alcohols refer to those that have at least six carbons; and the larger the number of carbons, the more hydrophobic the alcohol is. They are used as the basic oily ingredient for emulsification in cosmetics. Particularly cetanol is widely used in creams and can control the viscosity of the product by the amount added. The length of unsaturated alkyl chain affects the viscosity and feel of use. Oleyl alcohol, which has a double bond, is suitable for low viscous products and gives an oily and strong moisturized feeling. Branched-chain alcohols such as hexyldodecanol, isostearyl alcohol, and octyldodecanol cannot be used as main oily components for forming emulsions. However they can be used in creams combined with saturated alcohol for adjusting the viscosity and feel of use. Higher alcohols used in cosmetics are listed in Table 2.4.

2.1.5.1 Straight-Chain Alcohols

Lauryl alcohol: This C12 alcohol melts at 25°C, gives a distinct sweet smell, and is rarely used for cosmetics. When used to make nonviscous cream of light texture, it is combined with other straight-chain alcohols, such as cetanol.

Myristyl alcohol: This C14 alcohol melts at 35–42°C, is crystalline, and is almost odorless. It is used as a stabilizer, oily thickener, and emollient. Because it has fewer carbons than cetanol, it is more hydrophilic and gives a lighter and smoother feeling. It is suitable for leave-on skin and hair creams. It is more stable than lauryl alcohol and is almost odorless.

Cetanol is the best higher alcohol for making creams, particularly for increasing the viscosity. The viscosity, temperature stability, and feel of use of the cream also depend on the content of cetyl alcohol (C16) and the alkyl compositions of other ingredients. Cetanol of 7:3 in C16:C18 is widely used. There is 99% C16 cetanol but it cannot be used alone because it turns into liquid at around the melting point and destabilizes creams. Use of C18 together with C16 results in wider alkyl distribution and better stability and resistance against deterioration and low and high temperature stresses.

Stearyl alcohol is mainly C18 and is used like cetanol. The content of stearyl alcohol determines the viscosity, temperature stability, and feel of use of the cream.

Arachidyl alcohol: This C20 higher alcohol is rare and is scarcely used in cosmetics except for some hair wax products.

Behenyl alcohol has six more carbon atoms than cetanol, and thus has a higher melting point and is less polar. When the same amount is added, behenyl alcohol produces less viscous creams than cetanol and stearyl alcohol. It is useful for giving a soft and persistent texture. It can also be used for adjusting the hardness of solid products that do not contain water. Because it has an OH group at the end of the alkyl group, it is compatible with low-HLB nonionic surfactants and stabilizes the products.

Cetostearyl alcohol is a mixture of about 50% cetanol (C16) and 50% stearyl alcohol (C18). It is used like cetanol and stearyl alcohol.

2.1.5.2 Unsaturated Alcohol

Oleyl alcohol (C18) has a double bond at the center of the alkyl group. It does not increase the viscosity of the cream when used alone. When combined with cetanol, it can produce soft cream that feels very oily and moisturizing.

2.1.5.3 Branched-Chain Alcohols

Hexyl decanol is a C16 alcohol and has a short branched alkyl chain. It is rarely used in cosmetics. When combined in creams, it results in low viscosity. Addition of a large amount may spoil high temperature stability. The touch is light.

Isostearyl alcohol is a C18 alcohol and has a branched alkyl chain. It gives a smooth and silky feeling when combined in creams. It is suitable for leave-on hair cosmetics.

Octyldodecanol is a C20 alcohol and has a branched alkyl chain. It is widely combined in creams. It gives a very smooth texture and is suitable for both rinse-off and leave-on type cosmetics.

Sterols: Cholesterol and phytosterol have sterol frameworks. Small amounts are used for giving a moisturized feeling and for helping emulsification.

Table 2.4 Higher alcohols used in cosmetics

Alkyl group	Higher alcohol	Molecular formula	Molecular weight	Specific gravity	Melting point
n-6	Caproyl alcohol	$C_6H_{14}O$	102.18	0.8191	–44.6
n-8	Caprylyl alcohol	$C_8H_{18}O$	130.23	0.8254	–14.9
n-10	Caprylic alcohol	$C_{10}H_{22}O$	158.29	0.8297	6.9
n-12	Lauryl alcohol	$C_{12}H_{26}O$	186.34	0.8333	24.0
n-14	Myristyl alcohol	$C_{14}H_{32}O$	214.39	0.8355	37.9
n-16	Cetyl alcohol	$C_{16}H_{34}O$	242.42	0.8375	49.3
n-18	Stearyl alcohol	$C_{18}H_{38}O$	270.50	0.8392	58.0
n-20	Arachidyl alcohol	$C_{20}H_{42}O$	298.56	0.8405	65.6
n-22	Behenyl alcohol	$C_{22}H_{46}O$	326.61		70.6
e-18′	Oleyl alcohol	$C_{18}H_{26}O$	268.49	0.8485	2.0
iso-16	Hexyldecanol	$C_{16}H_{34}O$	242.00		
iso-18	Isostearyl alcohol	$C_{18}H_{28}O$	270.00		
iso-20	Octyldodecanol	$C_{20}H_{42}O$	298.00	0.84	

2.1.6 *Fatty Acids*

Fatty acids have a COOH group at one of its terminals and are thus highly polar. Because most of the commercially available ingredients for cosmetics contain fatty acids, it is important to know their alkyl compositions and choose suitable fatty acids for the purpose (check Tables 2.5 and 2.6). Because fatty acids react with other substances and may undergo saponification by alkalis or form complexes with cationic surfactants, the reactivity in formulation should also be examined. The high reactivity can be used for forming complexes and improving the feel of use.

Lauric acid: This C12 straight-chain fatty acid melts at 28°C. It is used for manufacturing liquid soap cleansers and body soap products by neutralizing with potassium hydroxide in advance.

Myristic acid: This C14 straight-chain fatty acid melts at 28°C. It is used together with lauric acid for manufacturing liquid body soap products by neutralizing with potassium hydroxide in advance. It is also used to produce soap creams by mixing with palmitic acid and steric acid and neutralizing with potassium hydroxide.

Palmitic acid: This C16 straight-chain fatty acid is usually available in mixtures with stearic acid.

Stearic acid: Mixtures of stearic acid and palmitic acid are the most widely used fatty acids in cosmetics, particularly creams. There are commercially available mixtures of various stearic to palmitic acid ratios, and those suitable for the purpose should be chosen. When used as the emulsifier of creams, high C18 content results in hard creams and coarse appearance. High C16 content produces smooth creams, but it is difficult to increase the viscosity.

Behenic acid: This C22 fatty acid has a high melting point and is difficult to neutralize and saponify with alkali. Due to the large alkyl group, it is rather hydrophobic. It is rarely used in cosmetics.

Oleic acid is an unsaturated fatty acid and has a double bond. It melts at 4°C and is easily oxidized. It is used in toilet soaps, liquid soaps, and cleansers. Combined with coconut oil fatty acids and lauric acids, it makes a suitable foaming and viscosity agent.

Isostearic acid is a branched-chain liquid fatty acid and has better oxidation stability than oleic acid. It is suitable for liquid cosmetics. It is stable against oxidization in the liquid state and is suitable as a substitute of oleic acid for producing low viscous creams. Combination with straight-chain alcohol such as cetanol is essential.

Fatty acids produced from coconut oil and palm-kernel oils consist of approximately 50% laurate, approximately 25% myristate, approximately 10% caprylic/capric, and some oleic acid and palmitic acid. They are suitable for producing liquid soap by neutralizing with potassium hydroxide, which foams well and has strong detergency. They are also used for creamy cleansers.

Table 2.5 Fatty acids used in cosmetics

Alkyl group	Fatty acid	Molecular formula	Molecular weight	Specific gravity	Melting point	Neutralization number
n-6	Caproic acid	$C_6H_{12}O_2$	116	0.93	–3.0	484
n-8	Caprylic acid	$C_8H_{16}O_2$	144	0.91	16.0	390
n-10	Caprylic acid	$C_{10}H_{20}O_2$	172			326
n-12	Lauric acid	$C_{12}H_{24}O_2$	200.3	0.8960(d60)	44.2	280.1
n-14	Myristic acid	$C_{14}H_{28}O_2$	228.4	0.8622(d54)	53.9	245.7
n-16	Palmitic acid	$C_{16}H_{32}O_2$	256.4	0.8487(d70)	63.1	218.8
n-18	Stearic acid	$C_{18}H_{36}O_2$	284.5	0.8390(d80)	69.6	197.2
n-20	Arachidic acid	$C_{20}H_{40}O_2$	312.52		75.3	180
n-22	Behenic acid	$C_{22}H_{44}O_2$	340.6	0.82212(d100)	81.5	164.7
e-18′	Oleic aid	$C_{18}H_{34}O_2$	282.5	0.8905(d20)	13.4	198.6
i-18	Linoleic acid	$C_{18}H_{32}O_2$	280	0.90	–18.0	200
	Linolenic acid	$C_{18}H_{30}O_2$	278			
iso-18	Isostearic acid	$C_{18}H_{36}O_2$	284.5	0.89(d20)	15 or lower	198
oh-18	Hydroxystearic acid	$C_{18}H_{36}O_2$	300.5		76–77	186.7
	Undecylenic acid	$C_{11}H_{20}O_2$	184.3	0.9072(d25)	24.5	304.5
e-14′	Myristoleic acid	$C_{16}H_{30}O_2$				
e-16′	Palmitoleic acid	$C_{18}H_{34}O_2$				

Table 2.6 Combinations of fatty acids and other ingredients and purpose of use

Alkyl group	Ingredients to combine					
of fatty acid	Anionic surfactant	Cationic surfactant	Nonionic surfactant	Fatty acids/higher alcohols	Feel of use	Stability
n-12	For liquids (shampoos)	For liquids	For liquids and solubilization	Liquids	Coarse	Increases smell
n-14				Liquids, emulsification	Smooth, lubricating	Stable
n-16	For creams	For creams and liquids	For creams	Creams, emulsification	Moisturized	Stable
n-18	For creams	For creams and liquids	For creams	Creams, emulsification	Moisturized	Stable
n-22		For creams (weak emulsification capacity)		Hard creams	Soft	Stable
e-18'			For low viscosity and solubilization	Liquids, emulsification	Oily, moisturized	Prone to oxidization
iso-16				Adjusting viscosity and texture	Light, lubricating	Stable
iso-18				For creams and liquids	Lubricating	Stable
iso-20				For creams and liquids	Lubricating	Stable
oh-18				Gelatinization of oils		

2.1.7 Silicones

Silicones have a basic framework consisting of dimethylsiloxane, which is either circular or a straight chain. There are various forms such as monomers to which an amino group, alkyl group, POE or POP is introduced to a silica or carbon atom, copolymers and block polymers; and the variations are likely to increase further.

Dimethicone is the polymer that forms the basic framework of silicone. The degree of polymerization is expressed with viscosity, which ranges from 6 cs to several million in materials for cosmetics. Low viscous dimethylsiloxanes of 6–10 cs improve the vanishing characteristics (phenomenon of becoming white when rubbed by hand) of creams without influencing the texture. Those of 20–100 cs are oily and used for hair care products to increase lubrication. Dimethicones of 100–5,000 cs give a strong silky and smooth feeling. Those of 5,000 cs to tens of thousands cs give a persistent and silky smooth texture. Dimethylsiloxanes larger than 10,000 cs are highly viscous and are little compatible with other oily ingredients. They do not disperse uniformly and do not emulsify thoroughly. Therefore, there are premix products available, which are prepared by dissolving high-molecular-weight dimethicone in low-molecular-weight dimethicone and cyclomethicone, and are widely used for hair care products.

Cyclomethicone (dimethyl cyclopentasiloxane) in the pentacyclic form is mainly used in cosmetics. It is used as a solvent of highly polymerized methylpolysiloxane and widely in skincare and hair care products. It gives a light and smooth texture. Due to its low boiling point (210°C), it volatilizes under heat, and methods should be carefully designed for adding cyclomethicone when producing cosmetics. Because it gives a very light touch, it is suitable for leave-on cosmetics.

POE-POP dimethicones are nonionic surfactants consisting of dimethicone and POE-POP. They come in copolymers in which POE-POP group is bound at an end of dimethicone or in which POE-POP is added to a methyl group of the dimethyl end.

They are used as nonionic surfactants for emulsifying creams and silicone gels. Their hydrophilic and hydrophobic properties, which are determined by the molar number of the POE-POP component, are used for improving the feel of use of shampoos and the silky feeling of water-base cosmetics.

Amino-modified silicones have a structure in which an alkylamino group or its derivative is added to a methyl group of dimethicone. They are widely used for hair care products to give a good texture. The oily type is suitable for creamy conditioners. There is also a hydrophilic type, which is modified with polyoxyethylene and polyoxypropylene and is used for liquid cosmetics such as shampoos.

2.2 Surfactants

Surfactants have both hydrophilic and lipophilic groups in a single molecule. Most surfactants have a lipophilic alkyl group consisting of 12–22 carbon atoms (C12–C22). Surfactants are classified by the ionization state (anionic, cationic, nonionic, and amphoteric) of the hydrophilic group. This chapter outlines the properties of each kind of surfactant and then describes the characteristics of major surfactants used in cosmetics.

2.2.1 Anionic Surfactants

Anionic surfactants lather well and have high detergency, emulsification capability, and permeability. They are mainly used as the chief ingredient of shampoos, body soaps, and facial cleansers. Of those used for shampoos, many are produced from coconut oil and palm kernel oil and have an alkyl group consisting of C12 lauryl groups. Some of the surfactants are also used as emulsifiers as well as for shampoos. For emulsifiers, cetyl and stearyl groups are widely used.

Anionic surfactants, which have one hydrophilic and one lipophilic group, are easy to understand when they are classified into four by the hydrophilic group: sulfonic acid, sulfuric acid, carboxylic acid, and phosphate.

2.2.1.1 Structure and Properties

Hydrophilic Group and Properties

The most anionic hydrophilic group is those that have six oxygen and sulfur atoms combined. Phosphorus atoms have five valence electrons and are thus weaker than sulfur atoms in attracting electrons and are less anionic. Carboxyl groups have only two oxygen atoms and are little anionic. Therefore, sulfuric-acid surfactants are most anionic followed by sulfonic-acid, phosphate and carboxylic-acid surfactants in this order.

Particularly in shampoos, the complexes formed by surfactants and cationic polymers determine the viscosity and feel of use of the product. At low pHs, surfactants lose the surface activation capacity; but the stronger the anionic is the surfactant, the less it is affected. For example, amino acid-based anionic surfactants that have carboxyl groups require pH 6.5 or higher. On the other hand, anionic surfactants that have alkyl sulfate or sulfonate lather well and maintain the surface activation performance event at low pH (Fig. 2.1).

The effects of salts increase proportionally to the electron density of the hydrophilic group. The sulfuric-acid and sulfonic-acid surfactants, which have oxygen

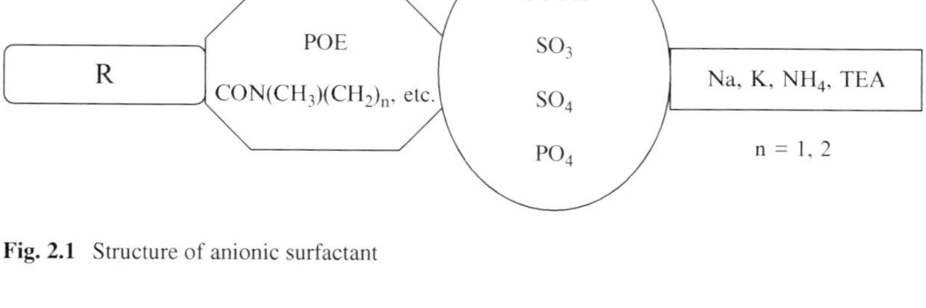

Fig. 2.1 Structure of anionic surfactant

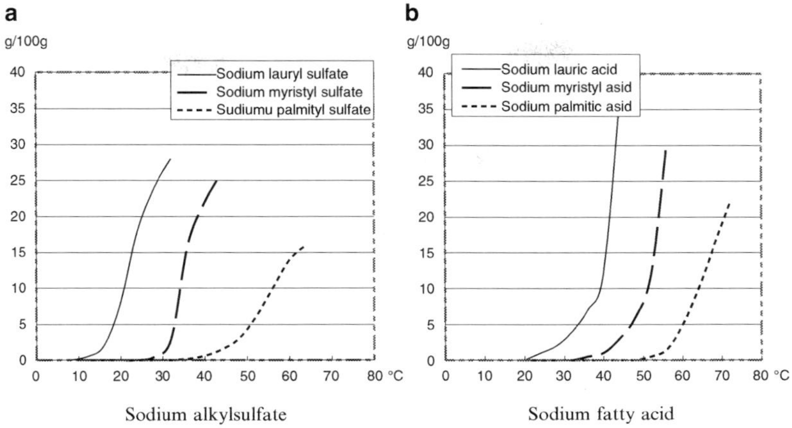

Fig. 2.2 Solubility of anionic surfactants [6]

and sulfur atoms, are little prone to salting out, but those that have carboxyl groups become cloudy when the salt content is high in the formulation of the shampoo.

Lipophilic Group and Properties

In most surfactants, the lipophilic groups are produced from coconut oil and palm kernel oil. Few surfactants have myristyl, cetyl, stearyl, and oleyl lipophilic groups.

The water solubility of a surfactant is correlated with the structure of the lipophilic group [2]. Figure 2.2 plots the water solubility of sodium fatty acid and sodium alkylsulfate surfactants against temperature. As the figure shows, surfactants that have longer alkyl groups require higher temperature to dissolve in water.

To achieve high detergency and lathering, lauryl and cocoyl are used since they have short alkyl groups of C12 and thus have high affinity to water. They easily take a bi-membrane structure and lather well. Longer alkyl groups of C16–18 are less water affinitive and more affinitive to oily ingredients, such as cetanol. Therefore, cetyl (C16) and stearyl (C18) are suitable for emulsification. Myristyl (C14) has intermediate properties and can be used for both purposes. Oleyl (C18) is suitable for emulsification and also for preparing transparent liquid because the molecule, which has a double bond, takes not a linear structure but a *cis*- or *trans*-configuration; and the three-dimensional structure hinders crystallization.

Improving the Properties by Group Insertion

The basic structure of anionic surfactant consists of a hydrophilic group and a lipophilic group. The undesirable properties attributable to the structure can be modified by inserting polyoxyethylene, amino acids, or a $CON(CH_3)C_2H_4$ group in between.

For example, insertion of polyoxyethylene increases hydrophilicity, which prevents separation at low temperatures. Insertion of $CON(CH_3)C_2H_4$ or $CON(H)CH_2$, which contain an amide group of CONR, increases softness and smoothness, reduces coarseness, and improves the feel of use. Sodium methyl cocoyl taurate, which has $CON(CH_3)C_2H_4$ between its alkylsulfonate groups, is more sensitive to salts than sodium alkylsulfonate due to the nitrogen atom and is prone to turbidity.

Counter Ions and Properties

Counter ions also affect the properties, solubility and transparency of surfactants. Widely used counter ions are sodium salts, but they can also be potassium, ammonium or triethanolamine salts. Sodium salts are most effective in lowering Krafft point and increasing the solubility, followed by potassium salt, ammonium salts, and triethanolamine, in this order.

Most sodium salts, potassium salts and ammonium salts precipitate at low temperatures and are thus not suitable for transparent products, while triethanolamine is suitable for transparent preparations. A 2% laurylsulfate sodium or laurylsulfate ammonium solution becomes cloudy at 17°C, while a 2% laurylsulfate triethanolamine solution is transparent even at –5°C. At 5°C, the triethanolamine solution is transparent, the sodium solution is white gel, and the ammonium solution is whitish liquid (Table 2.7).

2.2.1.2 Alkylsulfate and Polyoxyethylene

Alkylsulfate Salts

Alkylsulfate salts have an $R\text{-}O\text{-}SO_3M$ structure, and polyoxyethylene alkylsulfate salts are $R\text{-}O\text{-}(CH_2CH_2O)_n\text{-}SO_3M$. The latter has polyoxyethylene inserted between the alkyl and sulfate groups.

Alkylsulfate salts precipitate at low temperatures due to low Krafft point and low solubility. The insertion of polyoxyethylene between the alkyl and sulfate groups improves the affinity to the water, raises Krafft point, and prevents precipitation at low temperatures.

Sodium Lauryl Sulfate

$C_{12}H_{25}\text{-}O\text{-}SO_3Na$

Sodium lauryl sulfate is white powder of low solubility. It is unsuitable for transparent shampoos because it precipitates, becomes cloudy, and deposits at low temperatures. It may be included in pearl shampoos in small quantities. Because it is a sulfate salt, it is not affected by oxidation–reduction agents and can be used as an emulsifier of permanent wave solvents and anionic hair dye creams.

Ammonium Lauryl Sulfate

$C_{12}H_{25}$-O-SO_3NH_4

Ammonium lauryl sulfate is used in shampoos, but produces ammonia gas at high pHs because ammonia becomes counter ions. Therefore, the pH should be kept low.

Triethanolamine Lauryl Sulfate

$C_{12}H_{25}$-O-$SO_3N(C_2H_4O)_3$

This surfactant can be but not widely used for transparent shampoos because it causes coloring due to triethanolamine counter ions.

Sodium Cetyl Sulfate

$C_{16}H_{33}$-O-SO_3Na

Sodium cetyl sulfate has a longer alkyl group compared to sodium lauryl sulfate, and therefore is a suitable emulsifier for anionic creams. It gives a slightly more lubricative touch than sodium lauryl sulfate.

Polyoxyethylene Sodium Lauryl Ether Sulfate (INCI: Sodium C10-15 Pareth Sulfate)

$C_{12}H_{25}$-O-$(CH_2CH_2O)_nSO_3Na$

This anionic surfactant is widely used in transparent and pearl shampoos. It has superior low-temperature stability due to the insertion of 2 or 3 mols of polyoxyethylene between the alkyl and sulfate groups, and does not become cloudy even at low temperatures. It lathers well and is stable even in slightly acidic condition. Because it resists acids, alkalis, and oxidation–reduction agents, it is used as an emulsifier of permanent wave solvents and anionic hair dye creams.

Polyoxyethylene Ammonium Lauryl Ether Sulfate (INCI: Ammonium Laureth Sulfate)

$C_{12}H_{25}$-O-(CH_2CH_2O)n SO_3 NH_4

This surfactant is usually used together with ammonium lauryl sulfate and is suitable for low-pH products. It is rarely used in Japan.

2.2.1.3 PEG Fatty Acid Amide MEA Sulfate

$$\begin{array}{l}\ \ \ \ O\\ \ \ \ \ \|\\ R\text{-}C\text{-}NHCH_2CH_2O(CH_2CH_2O)\text{-}SO_3Na\end{array}$$

This anionic surfactant has a sulfate structure, in which monoethanol amide is bound to POE alkyl ether. When used in shampoos, it improves lathering and gives a favorable and heavy touch. Sodium PEG-4-cocamide sulfate is a typical surfactant of this group.

2.2.1.4 Alkyl Methyltaurate Sodium Salts

$$\begin{array}{l}\ \ \ \ O\ CH_3\\ \ \ \ \ \|\ \ \ |\\ R\text{-}C\text{-}NCH_2CH_2\text{-}SO_3Na\end{array}$$

These sodium salts have methyltaurine ($N(CH_3)CH_2CH_2$) between the alkyl and sulfonic acid groups of sodium lauryl sulfate. The alkyl group is cocoyl, lauryl, myristyl, cetyl, stearoyl, or oleoyl. The insertion of methyltaurine gives a soft, moderate and moist touch, and a better feel of use than lauryl acid salt.

Those that have cetyl, stearyl, or oleyl as the alkyl group are used as an emulsifier of creams.

Sodium cocoyl methyltaurine and sodium lauroyl methyltaurate can be used for shampoos, but salts should be removed for the use in transparent shampoos because they produce cloudiness and precipitation at low temperatures. Because the hydrophilic group is sulfate, they can produce transparent shampoos at pH 5.5 or lower. Sodium cocoyl methyltaurine gives a soft, smooth, and moderately moist feeling.

2.2.1.5 Alkyl Methylalanine Sodium Salts

$$\begin{array}{l}\ \ \ \ O\ CH_3\\ \ \ \ \ \|\ \ \ |\\ R\text{-}C\text{-}NCH_2CH_2\text{-}COONa\end{array}$$

This group of surfactants has a structure in which the sulfonate of alkyl methyltaurine is replaced by a carboxyl group. Sodium cocoyl methylalanine is more moisturizing and persistent than sodium cocoyl methyltaurine and a lighter and smoother feeling than TEA cocoyl glutaminate.

Sodium lauroyl methylalanine sodium and sodium cocoyl methylalanine can be used for transparent shampoos at pH6 or lower.

2.2.1.6 Alkyl Sarcosinate

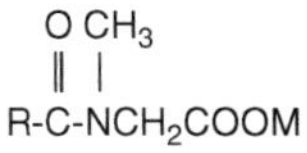

This group of carboxylic acid surfactant has one less carbon than the taurine group of alkyl methylalanine salts. The required pH of the shampoos is higher than for methyltaurine but is equivalent to that for alkyl methylalanine salts. It gives a slightly less moisturized feeling than sodium cocoyl methylalanine.

2.2.1.7 Olefin Sulfonate

This group of anionic surfactant has a structure of R-SO_3Na, lathers well and is hypoallergenic. They are stable at wide pH and temperature ranges and against acids, alkalis, and oxidation–reduction agents. When combined in shampoos with other anionic surfactants, they give a favorable and smooth feeling. They can also be used as an emulsifier of creamy preparations. Because they are stable against acids, alkalis and oxidation–reduction agents, they can be used in transparent permanent wave solvents, creamy preparations, and creamy hair dyes.

2.2.1.8 Alkyl Succinate

Disodium laurete sulfosuccinate

$R\text{-}COCH(SO_3Na)CH_2COONa$

Disodium laureth sulfosuccinate

$R\text{-}(CH_2CH_2O)_nCOCH(SO_3Na)CH_2COONa$

Dioctyl sodium sulfosuccinate

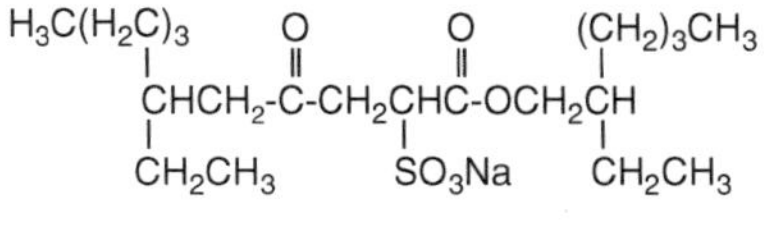

Disodium PEG-5 lauramide sulfosuccinate

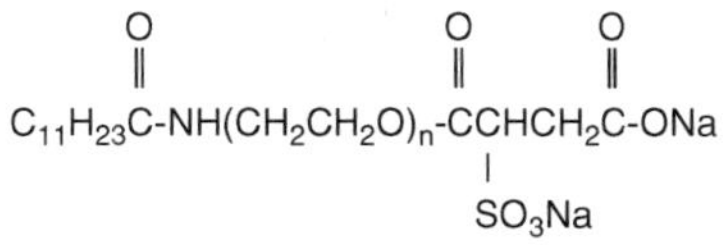

Alkyl succinates have one sulfonic acid and one carboxylic acid (with one or two alkyl groups) added to succinic acid. Penetration is good, but they rarely form micelles and do not much increase viscosity. They have rather low lathering capacity but are not irritating. They give a lubricative and favorable touch to shampoos when combined with sodium alkyl (C14–16) sulfonate or sodium polyoxyethylene laurylether sulfate.

2.2.1.9 Alkyl Phosphates

Lauryl phosphate

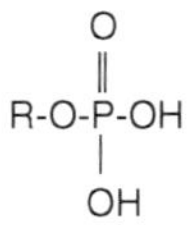

Cetyl phosphate

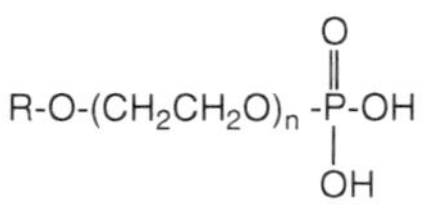

There are alkyl phosphates that have polyoxyethylene, dialkyl group, etc. They are usually neutralized with alkali prior to use to produce preparations of an arbitrary pH. They are used as detergents in body soaps and emulsifiers in cream formulation. They can also be combined with cationic surfactants and used for liquid emulsification and other various methods.

2.2.1.10 Fatty Acid Salts

Sodium Fatty Acids

They have a structure of R-COONa and are solid at the room temperature. Sodium fatty acids used for toilet soaps are mainly sodium salts of palmitic acid, oleic acid, and lauric acid.

Potassium Fatty Acids

They have a structure of R-COOK and are suitable for liquid soaps. To achieve stable liquid properties, those that are produced from C12 oils, such as coconut oil and palm kernel oil, need to be used. Some oleic acid can also be added. Addition of palmitic acid and stearic acid in large quantities results in cloudiness and deposition at low temperatures. Oleic acid potassium becomes pasty. The favorable pH range is from 10 to 10.5. At low pHs, they are prone to precipitation at low temperatures and chronologic oxidization and odor development. They lather well and are strongly detergent.

Polyoxyethylene Lauryl Ether Sodium Acetate

This surfactant has a structure of R-O-$(CH_2CH_2O)_n$-CH_2-COONa, which has polyoxyethylene inserted between the alkyl and carboxyl groups of fatty acid salts to prevent cloudiness and deposition at low temperatures. They are used in shampoos and as emulsifiers of creams.

2.2.1.11 Acylamino Acid Salts

Fatty Acyl-L-Glutamic Acid Salts

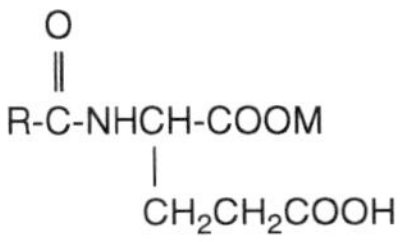

This anionic surfactant of the acyl amino acid type has fatty acid as the lipophilic group and amino acid as the hydrophilic group. The properties of this anionic surfactant group change greatly by the counter ions used. When the counter ions are sodium hydroxide or potassium hydroxide, the compound is solid and powder, which has low solubility and precipitates at low temperature, and is not suitable for liquid shampoos.

Those that have triethanolamine are suitable for liquid products. For transparent shampoos and other products, the pH should be kept at 6 or higher because the lipophilic groups are detached at low pHs, forming precipitation.

Fatty Acid Acyl-*N*-Aspartate

TEA lauroyl aspartate

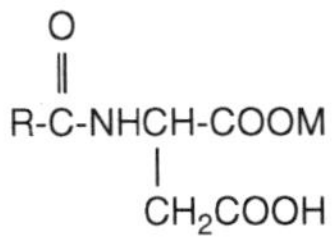

Sodium lauroyl aspartate has one less carbon than sodium lauroyl glutamate and is liquid despite it is sodium salt. It can be used for transparent shampoos.

Alkyl Methyl Alanine Salts

N-Sodium cocoyl alaninate

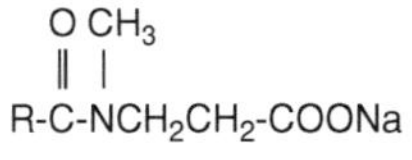

Sodium cocoyl alaninate can be used for transparent products at slightly lower pHs than anionic glutamate surfactants. Unlike glutamate sodium salts, which are powder, the alanine salts are liquid. They can produce transparent shampoos at pH 6 or lower by combining with other anionic surfactants. They give a moisturized and rather light feeling.

Alkyl Lactylate

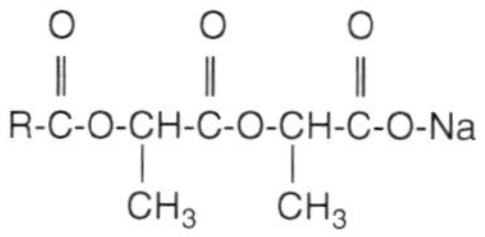

This group of anionic surfactants has lactic acid as the hydrophilic group and stearyl or isostearyl as the lipophilic group. Therefore, it is difficult to use in shampoos. It is suitable for emulsifiers of creams. Due to high safety, it is an approved food additive.

Alkyl Isethionate

This anionic surfactant has a structure of fatty acid ethyl ester sulfonate $RCOOCH_2CH_2SO_3M$. It has an ester bond and cannot be used in alkaline formulations because hydrolysis takes place. Due to its fatty acid ethyl ester component, it gives a dry and light feeling, and is more suitable than fatty acid salts for producing shampoos that give a refreshing feeling.

Table 2.7 Anionic surfactants

Alkyl sulfate and polyoxyethylene alkyl sulfate salts
Sodium lauryl sulfate
Ammonium lauryl sulfate
Triethanolamine lauryl sulfate
Sodium cetyl sulfate
Polyoxyethylene sodium lauryl ether sulfate (INCI: Sodium C10–15 pareth sulfate)
Polyoxyethylene ammonium lauryl ether sulfate (INCI: Ammonium laureth sulfate)
PEG fatty acid amide MEA sulfate
Sodium PEG-4-cocamide sulfate
Alkyl methyltaurate sodium salts
Sodium lauroyl methyltaurate
Sodium cocoyl methyltaurate
Sodium myristoyl methyltaurate
Sodium palmitoyl methyltaurate
Sodium stearoyl methyltaurate
Sodium oleoyl methyltaurate
Alkyl methylalanine sodium salts
Sodium lauroyl methyl alanine
Sodium cocoyl methyl alanine

(continued)

Table 2.7 (continued)

Potassium cocoyl methyl alanine
TEA lauroyl methyl alanine
Sodium myristoyl methyl alanine
Alkyl sarcosinate
Sodium cocoyl sarcosinate
Potassium cocoyl sarcosinate
Sodium lauroyl sarcosinate
Potassium lauroyl sarcosinate
TEA lauroyl sarcosinate
Olefin sulfonate
Sodium alkyl (C14–16) sulfonate
Sodium alkyl (C14–18) sulfonate
Alkyl succinate
Disodium Laurete silfosuccinate
Disodium laureth sulfosuccinate
Dioctyl sodium sulfosuccinate
Disodium PEG-5 lauramide sulfosuccinate
Disodium(C12–14) pareth sulfosuccinate
Disodium(C12–15) pareth sulfosuccinate
Alkyl phosphates
Lauryl phosphate
Cetyl phosphate
Stearyl phosphate
Dicetyl phosphate
Dioleoyl phosphate
POE lauryl ether phosphate
POE cetyl ether phosphate
POE oleoyl ether phosphate
Sodium dioleth-8 Phosphate
Di (C12–15) pareth-2 phosphate
Fatty acid salts
Polyoxyethylene lauryl ether sodium acetate
Acylamino acid salts
Fatty acyl-L-glutamic acid salts
TEA-Cocoyl glutamate
K-Cocoyl glutamate
Na-Cocoyl glutamate
Na-Stearyl glutamate
Fatty acid acyl-*N*-aspartate
TEA lauroyl aspartate
Alkyl methyl alanine salts
Sodium lauroyl methylalanine
Sodium cocoyl methylalanine
TEA cocoyl alanine
Alkyl isethionate
Sodium cocoyl isethionate
Ammonium cocoyl isethionate

2.2.2 *Cationic Surfactants*

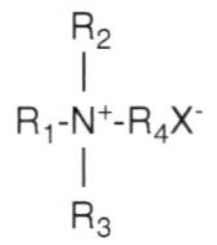

Cationic surfactants have a structure in which the four hydrogen atoms of an ammonium ion are replaced by alkyl and methyl groups. Changing one of the hydrogen atoms into an alkyl group of C12, C16, C18, or C22 and the remaining three into methyl groups and using chlorine as the counter ion forms most widely used mono-alkyl cationic surfactants, i.e. lauryl trimethyl ammonium chloride, cetyl trimethyl ammonium chloride, steartrimonium chloride, and behentrimonium chloride.

Besides their emulsification performances, cationic surfactants have efficacies not seen in other cosmetics ingredients such as disinfecting, making the hair flexible and antistatic, and giving a smooth touch to the skin and hair. This is possibly because the proteins that constitute hair contain more glutaminic acid and aspartic acid than basic amino acids and have rather anionic properties.

Because of the properties, cationic surfactants are widely used in hair care products, such as permanent wave lotions, quasi-drug class hair dyes, and leave-on and rinse-off hair conditioners.

2.2.2.1 Structure and Properties

Emulsification and Properties

There are mono-alkyl and di-alkyl cationic surfactants, depending on the number of alkyl groups. Mono-alkyl surfactants are effective emulsifiers and indispensable for hair care cosmetics. Di-alkyl cationic surfactants include dicocoyl, di-(C12–C18), dicetyl and distearyl surfactants. The di-alkyl surfactants are weak emulsifiers and cannot be used alone but need to be combined with mono-alkyl surfactants to emulsify ingredients into creams. They can provide various feels of use, which depend on the kind and amount of mono-alkyl surfactant combined.

Possible counter ions include chlorine, bromine, methosulfate, and saccharinate. Chlorine results in strongest emulsification performance, while bromine makes a weaker emulsifier. Methosulfate is safe but is weak in emulsification. Those that have alkyl groups of C16 and C18 are strong emulsifiers. Longer alkyl groups give more moisturizing, smooth and remaining feeling likely because longer alkyl groups are more hydrophobic (Table 2.8).

2.2.2.2 Mono-Alkyl Cationic Surfactants

Laurtrimonium Chloride

This cationic surfactant has a short alkyl group C12 and is not suitable as an emulsifier. It is a suitable solubilizer for transparent systems because it does not precipitate at low temperatures. It is also an effective solubilizer for cetyl trimethyl ammonium chloride and can be used for liquid hair care products. It is also applicable for form-shaped cosmetics because it lathers well.

Cetrimonium Chloride

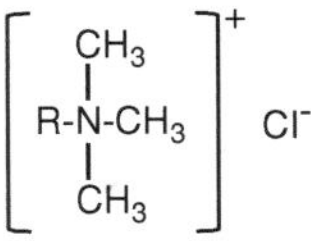

Having a hydrophilic group of cetyl (C16), this surfactant has strong emulsification capacity. It can improve the emulsification performance of behentrimonium chloride and behentrimonium methosulfate, which have big alkyl groups. It gives a light touch and is suitable mainly for rinse-off hair makeup products. It can be used for producing transparent liquid products, which do not become cloudy at low temperatures, by combining with lauryl trimethyl ammonium chloride.

Steartrimonium Chloride

Having stearyl (C18) as the alkyl group, steartrimonium chloride is a most effective emulsifier. It is the most widely used cationic surfactant for leave-on and rinse-off products. It has strong emulsification capacity and is suitable for cream conditioners. It gives a balanced moisturized and smooth feeling

Behentrimonium Chloride

Having a behenyl group, which consists of C22, behentrimonium chloride has weak emulsification capacity and cannot produce cationic creams by itself. Combination with a cationic or nonionic surfactant(s) is essential, such as cetrimonium chloride and steartrimonium chloride.

Due to its large alkyl group of C22, it is hydrophobic and tends to remain on the hair. Thus, it is widely used in professional-use rinse-off hair conditioners. It gives a strongly persisting, soft and very smooth touch. It is not suitable for leave-on hair care products because it gives a heavy feeling.

Cetrimonium Bromide

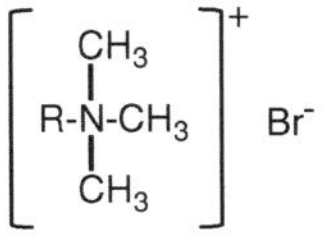

The chloride ion in cetrimonium chloride was replaced by bromine. It gives a similar feeling but is weaker in emulsification capacity than cetrimonium chloride. It is used for rinse-off hair conditioners.

Steartrimonium Bromide

The structure is the same as steartrimonium chloride except that the counter ion is bromine. It gives a smoother and softer touch than steartrimonium chloride and is used for professional-use hair treatments.

Hydroxyethyl Oleyl Dimonium Chloride

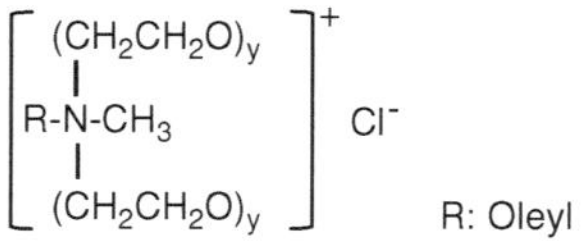

Because the oleyl group has a polyoxyethylene structure, it is stable at low temperatures and can thus be used for transparent liquid hair care products. It is used combined with steartrimonium chloride to make creams. Due to the oleyl group, it gives a moisturizing feeling. Because it gives a heavy touch in leave-on products, it is suitable for rinse-off products.

Behenyl PG-Trimonium Chloride

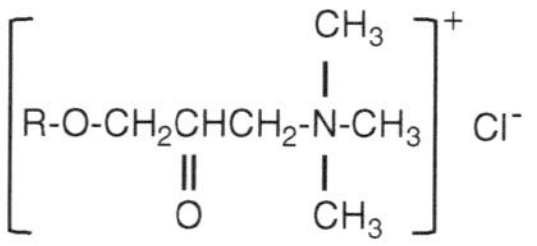

This cationic surfactant has improved emulsification performance compared to behentrimonium chloride. It gives a favorable soft and smooth feeling but a slightly too light touch due to its polyoxyethylene component.

Behentrimonium Methosulfate

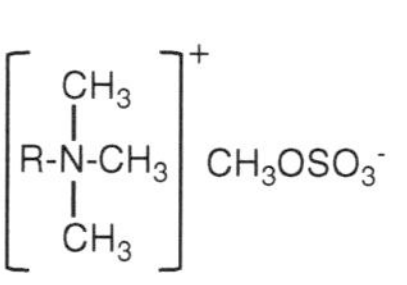

This cationic surfactant has slightly lower emulsification capacity than behentrimonium chloride. It is derived from lanolin oils and thus gives a favorable touch.

Stearyl Hydroxypropyl Trimonium Chloride

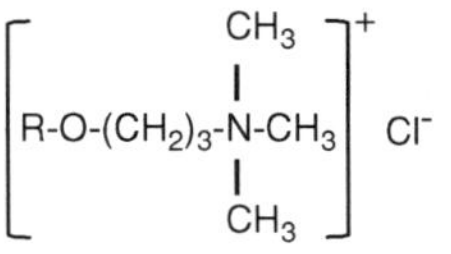

The structure is trimethyl ammonium consisting of hydroxypropyl and octadecanamide. It gives a slightly lighter touch than steartrimonium chloride.

Cetrimonium Saccharinate

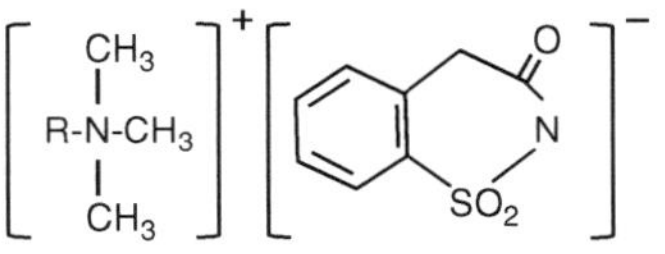

This surfactant has saccharin as the counter ion. It is unsuitable for creams and emulsive formulations but suitable for liquid preparations. It produces transparent viscous liquids, but becomes cloudy at low temperatures under the presence of salt. It gives a light and lubricious touch.

2.2.2.3 Di-alkyl Cationic Surfactants

Cocodimonium Chloride

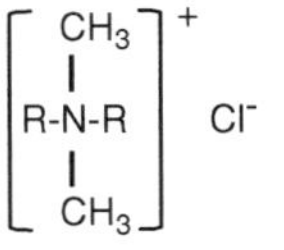

Cationic surfactants of the dialkyl type are not used as an emulsifier alone, but give a very good touch when combined with a cationic mono-alkyl surfactant. Cocodimonium chloride gives a strong lubricious feeling. Those that have longer cationic alkyl groups give stronger moisturized touch.

Increasing the quantity for emulsification lowers the viscosity of the cream. It cannot be used for transparent products because the dialkyl lauryl group causes cloudiness at low temperatures.

Di-C12–18 Alkyl Dimonium Chloride

The properties are similar to those of cocodimonium chloride.

Cetyl Dimonium Chloride

Because it is dicetyl, it cannot emulsify ingredients into creams by itself. It gives a strong smooth, soft, persistent and coated feeling, and can produce good hair conditioners when combined with steartrimonium chloride and behentrimonium chloride.

Stearyl Dimonium Chloride

This di-alkyl surfactant gives a smooth and persistent touch when combined with cationic surfactants of the alkyl or mono-alkyl type. Addition of a large quantity may reduce the lubricative feeling. Di-alkyl cationic surfactants that have longer di-alkyl groups (dicocoyl<dicetyl<distrearyl) give more persistent and less lubricative feeling.

Linole Amidopropyl Dimonium Ethosulfate

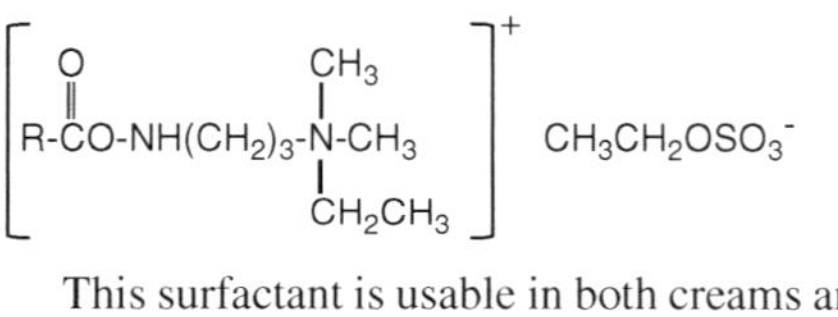

This surfactant is usable in both creams and transparence liquids but not alone. It is suitable for giving a wet and moisturized touch.

Dicocoethyl Hydroxyethylmonium Metholsulfate

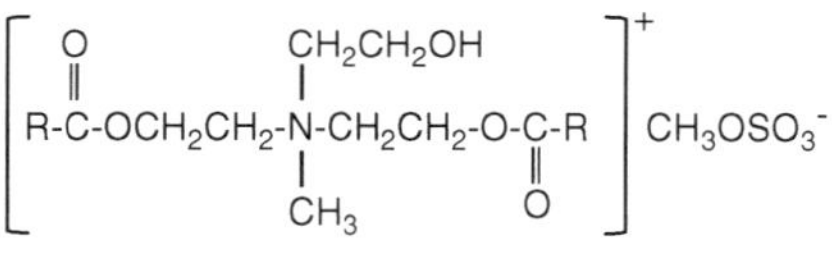

Dicocoethyl hydroxyethylmonium metholsulfate resembles cocodimonium chloride but gives a slightly lighter touch. Because it cannot emulsify ingredients by itself, it should be used together with other cationic surfactants. It produces depositions in transparent preparations at low temperatures.

Sodium Coco PG-Dimonium Chloride Phosphate

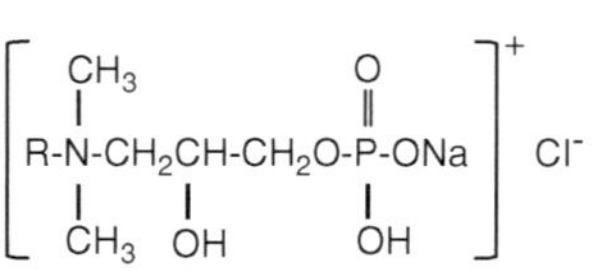

Sodium coco PG-dimonium chloride phosphate has a structure in which propylene glycol is added to coconut oil fatty acids and sodium phosphate as the cationic group. This very safe cationic surfactant is antibacterial and can be used in wipe-off type preparations to be soaked in nonwoven fabrics. It can be combined in shampoos to produce conditioner-in shampoos. It gives a less persistent and adsorptive touch than steartrimonium chloride and cetrimonium chloride, and gives a smooth feeling.

Linoleamidepropyl PG-Dimonium Chloride Phosphate

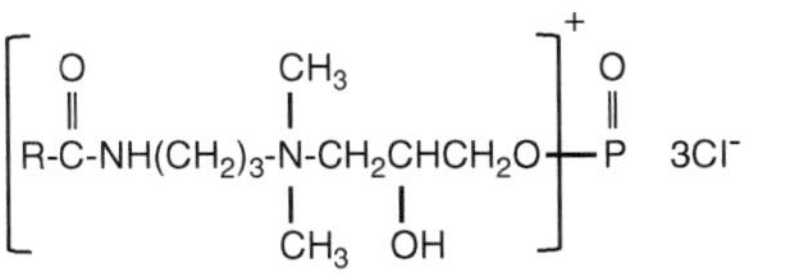

It has a structure in which the amide propyl group of linoleate or linolenic acid is cationized and propylene glycol is added. It is a safe cationic surfactant. It can be used for conditioner-in shampoos. It gives a light and lubricative touch. Combined with steartrimonium chloride to produce hair conditioners of the alkyl type, it gives a light and flexible touch.

2.2.2.4 Other Cationic Surfactants

Quaternium-91

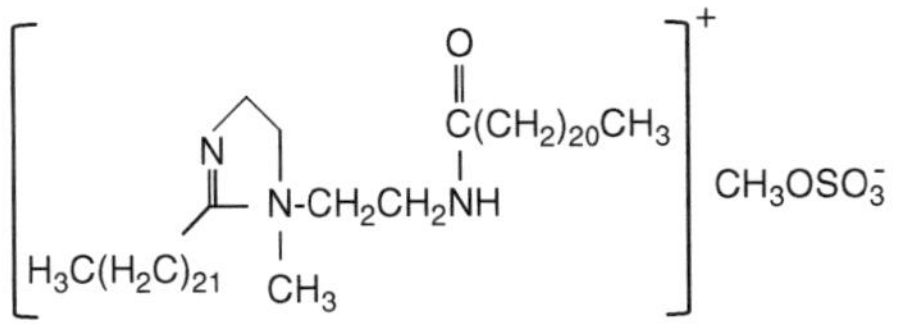

This methosulfate has an imidazoline structure that has two alkyl groups: behenyl and behenyl amide. It is an emulsifier for hair care products such as cream conditioners.

2.2.3 Tertiary Amines

There are three types of tertiary amines: alkyl amideamine, alkylamine and those that have polyoxyethylene added. They are used as surfactants by neutralizing with an acid, such as lactate, citric acid, phosphoric acid, glutaminic acid and stearic acid, which determines the feel of use. The pH can be adjusted by controlling the quantity of the acid, which also affects the viscosity and feel of use of the cream. This is because not-neutralized tertiary amines are highly polar cationic oils. When used in hair conditioners, it gives an inferior touch compared to fourth grade ammonium salt, but the feel of use can be improved by using them together.

Table 2.8 Cationic surfactants

Mono-alkyl surfactants
Laurtrimonium chloride
Cetrimonium chloride
Steartrimonium chloride
Behentrimonium chloride
Cetrimonium bromide
Steartrimonium bromide
Hydroxyethyl oleyl dimonium chloride
Behenyl PG-trimonium chloride
Behentrimonium methosulfate
Stearyl hydroxypropyl trimonium chloride
Cetrimonium saccharinate
Di-alkyl surfactants
Cocodimonium chloride
Di-C12-18 alkyl dimonium chloride
Cetyl dimonium chloride
Stearyl dimonium chloride
Linole amidopropyl dimonium ethosulfate
Dicocoethyl hydroxyethylmonium metholsulfate
Sodium coco PG-dimonium chloride phosphate
Jimo acid amide propyl aluminum chloride re-Dorin pg linoleic acid
Other cationic surfactants
Quaternium-91

2.2.3.1 Aminoalkylamides

Dimethylamino propyl stearamide

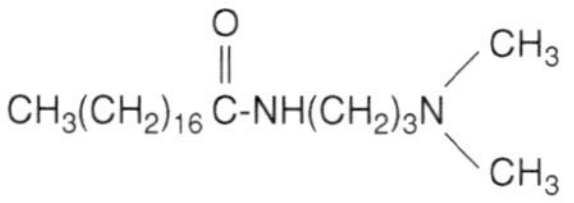

Behenamidopropyl dimethylamine

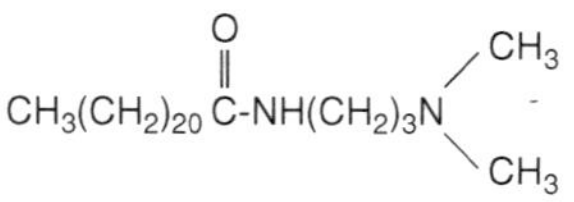

2.2.3.2 Alkyl Amines

Dimethyl stearamine

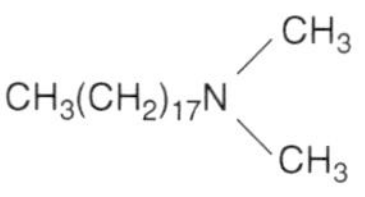

Behenyl dimethylamine

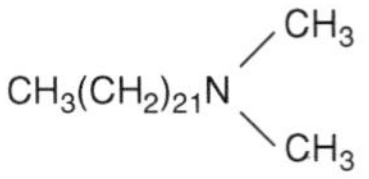

POE coconut alkylamine

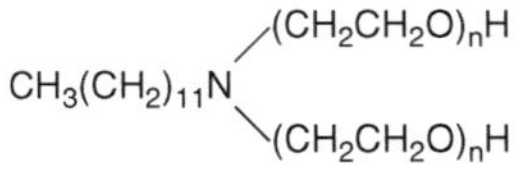

POE oleylamine

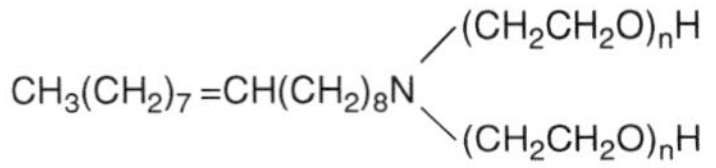

POE stearamine

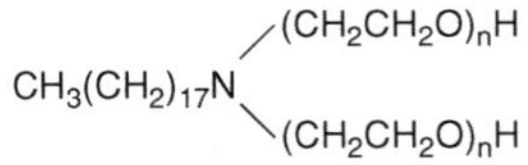

2.2.4 Amphoteric Surfactants

The basic structure of amphoteric surfactants consists of amino acids. Most amphoteric surfactants have an alkyl group consisting of C12. This type of surfactants is mainly used in shampoos and body soaps to improve detergency, lathering, and feel of use and reduce irritation. Although amphoteric surfactants have amino acids, which contain nitrogen atoms, they give a similar touch to that of anionic surfactants. They improve the feel of use by using together with anionic surfactants and cationic polymers (Table 2.9).

2.2.4.1 Alkyl Betaines

Lauryl dimethyl betaine

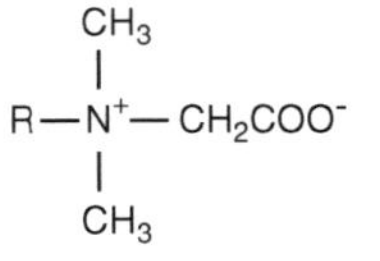

Coconut oil fatty acids and dodecyl dimethyl betaine improve the lathering of shampoos and give a light touch. Myristyl betaine and stearyl betaine are usable in creamy cleansers.

2.2.4.2 Amide Betaines

Lauric amide propyl betaine

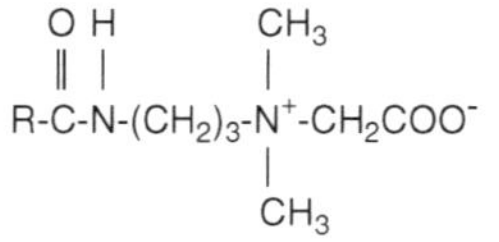

Coconut oil fatty acid and lauric amide propyl betaine are the most widely used amphoteric surfactants in shampoos. Combined with anionic surfactants and cationic polymers, it gives a favorable lubricative touch.

2.2.4.3 Sulfobetaines

Laurate amide propyl hydroxy sulfobetaine

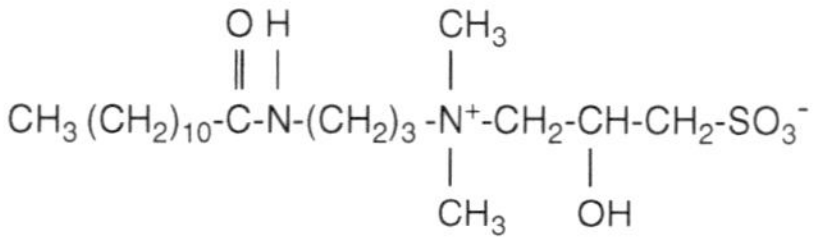

Lauryl hydroxy sulfobetaine

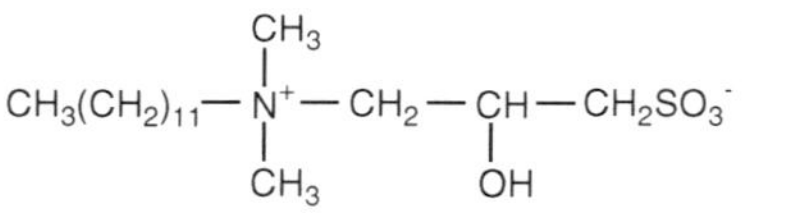

Laurate amide propyl hydroxy sulfobetaine is suitable for improving the lubricative touch. It is used in shampoos to make them light. It also lathers well.

2.2.4.4 Imidazolinium Betaines

N-alkyl-*N*′-carboxymethyl-*N*′-hydroxyethyl ethylenediamine sodium

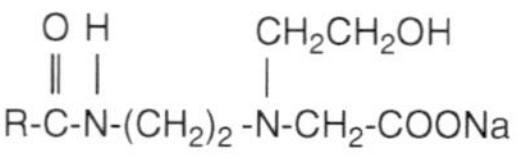

Disodium cocoamphodiacetate

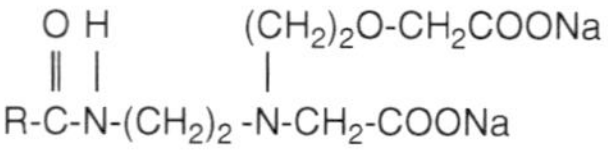

N-coconut oil fatty acid acyl-*N*-carboxymethyl-*N*-hydroxyethyl ethylenediamine sodium is little stimulating. It is suitable to reduce irritation. When used in shampoos, it gives a less sliding touch than cocamidopropyl betaine (amido betaine

surfactant) and gives a rather moisturized feeling. It is suitable for increasing the feel of persistence. It rather increases the viscosity when combined with POE sodium lauryl sulfate.

2.2.4.5 Propionic Acids

Sodium lauraminopropionate

$CH_3 (CH_2)_{11}$-NH-$(CH_2)_2COONa$

Disodium lauryliminodiacetate

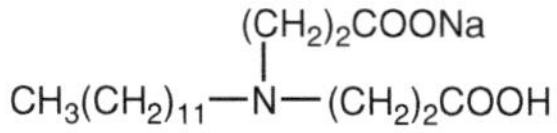

This group of amphoteric surfactants has the same structure as imidazolinium betaine surfactants but the anionic group is sodium propionate. When used in shampoos, it gives a lighter touch than imidazolinium betaine and lathers well.

2.2.4.6 Amine Oxides

Cocoamine oxide

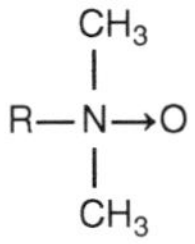

Lauramidopropylamine oxide

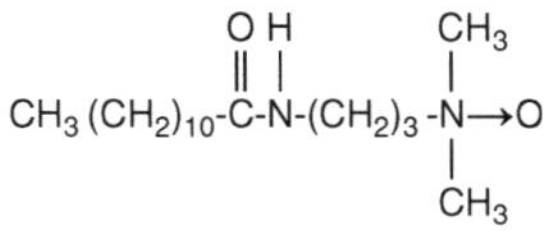

Amphoteric surfactants of the amine oxide type are used in shampoos to improve lathering and increase viscosity. It is better used together with amido betaine surfactants than alone to improve the lubricative touch of shampoos. There are lauryl amine oxide and lauramidopropyl dimethylamine oxide, which has an amide propyl group inserted between the alkyl group and amine oxide.

2.2.4.7 Other Amphoteric Surfactants

N-[3-Alkyl(12,14)oxy-2-hydroxypropyl]-L-arginine hydrochloride can be used as an emulsifier of cream-type hair conditioners. Combined with cationic surfactants, it gives a persistent feeling. It is unsuitable for shampoos. compound missing

Table 2.9 Amphoteric surfactants

Alkyl betaine
Coconut oil fatty acids (lauryl dimethyl betaine)
Cocobetaine
Lauryl betaine
Myristyl betaine
Stearyl betaine
Amido betaines
Coconut oil fatty acid
Lauric amide propyl betaine
Cocamidopropyl betaine
Palm kernel amidopropyl betaine
Lauric amide propyl betaine
Sulfobetains
Laurate amide propyl hydroxy sulfobetaine
Lauryl hydroxysulfobetain
Imidazolinium betaines
N-alkyl-*N*′-carboxymethyl-*N*′-hydroxyethyl ethylenediamine sodium
Sodium lauroamphoacetate
Sodium cocoamphoacetate
Disodium cocoamphodiacetate
Propionic acids
Lauramidopropylamine oxide
Lauramidopropyl dimethylamine oxide
Amine oxides
Lauramidopropyl dimethylamine oxide
Lauryl amine oxide
Other
N-[3-Alkyl(12,14)oxy-2-hydroxypropyl]-L-arginine hydrochloride

Hydroxylalkyl (C12–14) hydroxyethyl sarcosine

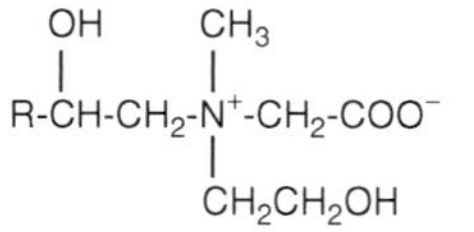

When used in shampoos, it gives a lighter touch.

2.2.5 Nonionic Surfactants

Nonionic surfactants that have lipophilic groups of C12–C18 are mainly used as emulsifiers. Those that consist of a small number of carbons are used for other special purposes. Most common hydrophilic groups are glycerin and polyethylene. The hydrophilic group can also be sorbitan, sorbitol, glucose and combinations of glycerin, polyethylene glycol, and polypropylene glycol. The bond between the lipophilic and hydrophilic groups is either ester or ether bond.

2.2.5.1 Structure and Properties

Nonionic surfactants, especially those that have polyethylene glycol as the hydrophilic group, show changes in apparent hydrophile–lipophile balance (HLB) by temperature. At a certain temperature, which is called the clouding point, the polyethylene glycol chain extends, loses hydrophilicity, and becomes hydrophobic; and the aqueous solution of polyoxyethylene (POE) nonionic surfactant becomes cloudy. This phenomenon is attributable to the increased distance between the oxygen atoms in the POE chain, which causes a drop in electron density.

The clouding point is higher in surfactants of higher PEO mols (i.e., longer POE chains) and is lower in lower mol POE compounds. Salts lower the apparent clouding points. Emulsions produced using only PEO nonionic surfactants are unstable to temperature changes.

Length of the Alkyl Group and Viscosity

The length of the alkyl group influences the properties of the products. From C12 to C18, the longer the alkyl group results in the higher viscosity of the products. However, oleyl (C18) produces low viscous products.

Cetyl (C16) and stearyl (C18) are suitable for emulsification. Lauryl (C12) is also suitable for emulsification, which results in lower viscosity than the former two, and also for solubilization. The viscosity and consistency of products can be adjusted by appropriately combining the alkyl groups (Table 2.10).

2.2.5.2 Monoglycerol Esters of Fatty Acids

Monoesters of stearic acid and glycerin have low HLB. Combined with other nonionic surfactants, they are used as emulsifiers of skin creams and also as oily ingredients. Glyceryl monostearate is mainly used.

2.2.5.3 Polyglycerol Esters of Fatty Acids

This group of ester-type nonionic surfactant has polyglycerin as the hydrophilic group and is strong against acids. Because the hydrophilic group is polyglycerin and not polyoxyethylene, it is stable to temperature changes and can produce temperature-stable creams. The alkyl group may be mono-, di-, tri-, penta-, and other forms of diverse kind of alkyl. The molar number of polyglycerin is high. It is widely used as emulsifiers. Those that have high HLB are also used as solubilizing agents.

2.2.5.4 Sorbitan and POE Sorbitan

Combinations of polyoxyethylene sorbitan fatty acid esters and sorbitan fatty acid esters are suitable emulsifiers of creams. Low-HLB sorbitan fatty acid esters are combined with polyoxyethylene sorbitan fatty acid esters. It is possible to adjust the viscosity of creams while not changing the oil ingredients by changing the alkyl group of the surfactant. For example, the viscosity of cream prepared by emulsifying with sorbitan stearate and polyoxyethylene sorbitan stearate can be lowered by changing a part of hydrophilic polyoxyethylene sorbitan stearate into polyoxyethylene sorbitan oleate.

2.2.5.5 Polyoxyethylene Sorbitol Tetraoleate

This group of nonionic surfactant has high emulsification power; and those of 40 mol and 60 mol are used to emulsify and disperse low-polar liquid paraffin and vegetable oils and fats, respectively. It is unsuitable for emulsifying viscous creams but is suitable for liquid preparations rich in oily ingredients, nonaqueous transparent cleansing oils and milky bath oils.

2.2.5.6 Polyoxyethylene Hydrogenated Castor Oil

Polyoxyethylene hydrogenated castor oils of large molar numbers (40 or 60 mols) are used as solubilizing agents particularly for liquid preparation such as lotions because they help maintain the oily touch. Those of small molar numbers can be used as moisturizing agents and hydrophilic oily ingredients. They are used as emulsifiers by combining with other nonionic surfactants but not with other polyoxyethylene hydrogenated castor oils.

2.2.5.7 Polyoxyethylene Hydrogenated Castor Oil PCA Isostearate

This nonionic surfactant has pyrrolidone added to one of the POE chains of polyoxyethylene hydrogenated castor oil and isostearic acid forming an ester bond with the other PEO chain. It is used like polyoxyethylene hydrogenated castor oils.

2.2.5.8 POE Alkyl Ethers

Polyoxyethylene stearic acid is used as an emulsifier mainly of creams. It is little influenced by the HLB of oily ingredients and can produce smooth creams by combining with low-HLB polyoxyethylene (40E.O.) stearate, which serves as the main emulsifier. The amount of the surfactant necessary for emulsification can be reduced by combining with sorbitan surfactants. The viscosity can be controlled by selecting the appropriate alkyl group.

2.2.5.9 Polyoxyethylene Glycerides

It has a structure in which polyoxyethylene is added to monoglyceryl fatty acid. It is used for emulsification, solubilization, and dispersion of powders.

2.2.5.10 Polyoxyethylene Alkyl Ethers

Because the hydrophilic and lipophilic groups are bound by ether bonds, this group of surfactants is not easily decomposed by acids and alkalis and thus can be used together with oxidation-reduction agents. It is widely used for permanent wave solvents and hair dyes. Many kinds are available, which differ in alkyl group and POE molar number, and are useful. It can be used to produce creams by combining with polyoxyethylene (20E.O.) cetyl ether and polyoxyethylene cetyl ether of higher or lower HLB. Viscosity adjustment of products is easily achieved by combining polyoxyethylene oleyl ether, polyoxyethylene isostearyl ether and polyoxyethylene cetyl, which have different alkyl groups. Polyoxyethylene alkyl ethers can also be used for formulations of a high salt content by replacing a part by polyoxyethylene lauryl ether, which has a short alkyl group, to slightly increase the HLB. Addition of a small amount of polyoxyethylene stearyl ether or polyoxyethylene behenyl ether is effective for increasing and changing the viscosity of creams.

2.2.5.11 Polyoxyethylene Polyoxypropylene Alkyl Ethers

This group of surfactant, which has a polyoxypropylene group in the molecule, is suitable for solubilizing, dispersing, and emulsifying oily ingredients. There are butyl, lauryl, cetyl, and stearyl ethers. Emulsification is difficult with this group of surfactant alone so it is better to combine with other polyoxyethylene alkyl ethers. Transparent preparations can also be produced by combining with a cationic surfactant to solubilize oily ingredients. Those of low HLBs can be used as solvents of oily ingredient and for transparent liquid formulations of low water content by combining with an oily solvent that contains no water and no nonionic surfactant of high HLB.

2.2.5.12 POE and POP Block Polymers

This group of nonionic pluronic surfactant has polyoxyethylene as the hydrophilic group and polyoxypropylene as the lipophilic group. They can be used for various purposes (for solvents, solubilizers, moisturizers, thickeners, gelatinizers, and touch improvers, etc.) by controlling the lengths of the hydrophilic and lipophilic chains. They are rarely used as emulsifiers. Those in which the molar numbers of the chains are small are used as solvents and moisturizers, while those of large chains are used as thickeners and gelatinizers. Those that have bigger polyoxypropylene than polyoxyethylene are effective for giving an oil-like moisturizing touch.

2.2.5.13 Alkanolamides

Alkanolamides are widely used as thickeners and bubble stabilizers of shampoos, body soaps, and creamy cleansers. Those that have lauric acid, coconut oil fatty acid, or palm kernel oil fatty acid as the alkyl group are used in shampoos; and those that have palmitic acid, oleic acid, or isostearic acid are used for creamy products. They can also be used for hair treatments and creamy cleansers.

Cocamide diethanolamide (DEA), palmitamide DEA, and lauramide DEA have been used as very suitable thickeners of shampoos. The characteristics of these and other alkanolamides are summarized in Table 2.11.

Table 2.10 Nonionic surfactants

Monoglycerol esters of fatty acids
Glyceryl caprylate
Glyceryl laurate
Glyceryl myristate
Glyceryl palmitate
Glyceryl stearate
Glyceryl oleate
Glyceryl isostearate
Polyglycerol esters of fatty acids
Polyglyceryl-n caprylate
Polyglyceryl-n laurate
Polyglyceryl-n dilaurate
Polyglyceryl-n myristate
Polyglyceryl-n stearate
Polyglyceryl-n distearate
Polyglyceryl-n tristearate
Polyglyceryl-n tetrastearate
Polyglyceryl-n pentastearate
Polyglyceryl-n decastearate
Polyglyceryl-n oleate
Polyglyceryl-n dioleate
Polyglyceryl-n trioleate
Polyglyceryl-n pentaoleate
Polyglyceryl-n decaoleate
Polyglyceryl-n isostearate
Polyglyceryl-n diisostearate
Sorbitans
Sorbitan coconut oil fatty acid
Sorbitan laurate
Sorbitan palmitate
Sorbitan stearate
Sorbitan oleate
Sorbitan isostearate
Sorbitan sesquioleate
Sorbitan sesquiisostearate
Sorbitan trioleate
Polyoxyethylene sorbitans
PEG-20 sorbitan cocoate
Polysorbate-20

(continued)

Table 2.10 (continued)

Polysorbate-40
Polysorbate-60
Polysorbate-65
Polysorbate-80
Polysorbate-85
PEG-2 sorbitan isostearate
Polyoxyethylene sorbitol tetraoleate
Sotbeth-n tetraoleate
Polyoxyethylene hydrogenated castor oil
PEG-n hydrogenated castor oil
Polyoxyethylene hydrogenated castor oil PCA isostearate
POE-n hydrogenated castor oil PCA isostearate
POE Alkyl ethers
PEG-n laurate
PEG-n myristate
PEG-n stearate
PEG-n oleate
PEG-n isostearate
Polyoxyethylene glycerides
PEG-n caprylic/capric glycerides
PEG-n glyceryl cocoate
PEG-n glyceryl stearate
PEG-n glyceryl oleate
PEG-n glyceryl isostearate
Polyoxyethylene alkyl ethers
Polyoxyethylene lauryl ether
Polyoxyethylene myristyl ether
Polyoxyethylene cetyl ether
Polyoxyethylene lauryl ether
Polyoxyethylene stearyl ether
Polyoxyethylene behenyl ether
Polyoxyethylene oleyl ether
Polyoxyethylene alkyl (C12-14) ether
Polyoxyethylene polyoxypropylene alkyl ethers
PPG-x Buteth-n
PPG-x Deceth-n
PPG-x Laureth-n
PPG-x Ceteth-n
PPG-x Steareth-n
POE and POP block polymers
PEG/PPG-x/y copolymer
Poloxamer n
Alkanolamides
Cocamide DEA
Palmitamide DEA
Lauramide DEA
Myristamide DEA

(continued)

Table 2.10 (continued)

Stearamide DEA
Isostearamide DEA
Oleamide DEA
Cocamide methyl MEA
Cocamide MEA
Stearamide MEA
Cocamide MIPA
Polyoxyethylene cocamide MEA
Polyoxypropylene cocamide MIPA
Sucrose fatty acid esters
Sucrose laurate
Sucrose dilaurate
Sucrose myristate
Sucrose palmitate
Sucrose hexapalmitate
Sucrose stearate
Sucrose oleate
Sucrose distearate
Sucrose hexaerucic acid
Sucrose pentaerucic acid
Alkylglucosides
Caprylyl glucoside
Decyl glucoside
Coco glucoside
Lauryl glucoside
Myristyl glucoside

2.2.5.14 Sucrose Fatty Acid Esters

These esterified nonionic surfactants consist of fatty acids and sucrose. They are widely used as food additives and as an emulsifier in cosmetics. They prefer slightly acidic conditions and are prone to disintegration at high pH ranges. They are unsuitable for cleansers, foaming agents, and penetration agents.

The HLB of sucrose fatty acid sucrose ester is determined by the alkyl groups of the fatty acids and the ratio between sucrose and ester contents. They are suitable emulsifiers for creams once the HLB is appropriately matched.

Unlike other emulsifiers, which are added into the oil phase, sucrose fatty acid esters, particularly those of large HLBs, hardly dissolve in oils and form lumps. They should be dispersed in glycols first and then added to the water phase.

2.2.5.15 Alkylglucosides

These surfactants consist of higher alcohols and glucose, and are used as thickeners of shampoos and cleansers, etc.

Table 2.11 Alkanoamide nonionic surfactants

Lauramide DEA	
Cocamide DEA	Conventionally used and very suitable thickeners of shampoos
Palmitamide DEA	
Lauramide DEA	
Myristamide DEA	Have large alkyl groups and thus lather poorly and are very rarely used for shampoos
Stearamide DEA	
Isostearamide DEA	
Oleamide DEA	
Cocamide methyl MEA	Useful thickener for transparent and pearl shampoos because it maintains the viscosity and does not solidify even at low temperatures
Cocamide MEA	Most widely used thickener in pearl shampoos. The thickening effect is milder than cocamide DEA
Stearamide MEA	Gives a smooth feeling to rinse off hair care creams, but has a high melting point and is little compatible to and hardly dissolves in oils
Cocamide MIPA	Can be used in pearl shampoos, but viscosity is harder to develop than cocamide MEA. Combining with POE cocamide MEA and polyoxypropylenecocamide MEA helps viscosity development
POE cocamide MEA	Those having polyoxyethylene chains of 2, 3, 5, and 10 mol are commercially available. The viscosity of shampoos is harder to develop than cocamide DEA
Polyoxypropylene cocamide MIPA	More effective thickener than polyoxyethylene cocamide MEA. Can be used for transparent shampoos

2.3 Polymers

Polymers are also widely used in cosmetics mainly for thickening and also for stabilizing, hair setting, and improving the feel of use.

The viscosity of polymers is determined by the molecular weight, and the chemical structures of the hydrophilic group and alkyl chain determine the properties. Even polymers of the same chemical name and structure may have different properties due to the differences in the conditions of synthesis and reaction. Polymers should be thoroughly tested before use because the properties, feel, stability, etc. may vary depending on manufacturer.

2.3.1 Structure of Polymers

Various polymers are used in cosmetics. They can be classified by the ionization state of the hydrophilic group into anionic, cationic, nonionic, and amphoteric polymers. Chemical constituents may be cellulose, acrylic acid, methacrylic acid, urethane, maleic acid, vinylpyrrolidone, dimethylaminopropyl methacrylic acid, acrylamide, dimethyldiallyl ammonium chloride, 2-hydroxy-3-(trimethylammonio) propylether chloride, propyl trimethylammonium chloride, etc.

To use polymers in cosmetics, care should be taken on the stability and compatibility with other ingredients (Table 2.12).

Table 2.12 Points to note for using polymers

(1) Always check for stability. Polymers may cause precipitation, coagulation and/or separation depending on combination with other ingredients event at same ionicity
(2) Low and high pHs may break bonds and reduce the viscosity of polymers
(3) Salts may also reduce viscosity

2.3.2 Purposes of Using Polymers

2.3.2.1 Thickening

Thickening is the most common objective of using polymers. Thickening polymers have cellulose or acrylate structures in general.

Anionic polymers for thickening aqueous systems include carboxyvinyl polymers, sodium carboxymethyl cellulose, xanthan gum, and polyacrylate (Table 2.13).

Nonionic polymers include, in the descending order of solubility to water, methylcellulose, hydroxyethylcellulose, hydroxypropyl methylcellulose, and hydroxypropylcellulose. Methylcellulose is highly hydrophilic and thus easily dissolves in cold and hot water. Hydroxyethylcellulose is little soluble to cold water and is dissolved by dispersing in cold water and then heating the dispersion. Hydroxypropylcellulose and hydroxypropyl methylcellulose are soluble to cold water and are thus dissolved by dispersing in hot water and then cooling the dispersion. They become hydrophobic, precipitate and become white solid at 42–45°C. The phenomenon occurs because of high hydrophobicity and can be prevented by combining large amounts of ethanol and glycol.

Polyquaternium

Cationic polymers used for thickening products include polyquaternium-10, polyquaternium-67, and guar hydroxypropyl trimonium chloride. They can be used combined with cationic surfactants, and are used for thickening and improving the feel of use of hair care products. Ingredients called polyquaterniums are summarized in Table 2.14.

Table 2.13 Polymers for thickening

Carbomer	Carboxyvinyl polymer
Acrylic acid/(C10-30) alkyl acrylate copolymer	Acrylic acid/(C10–30) alkyl acrylate copolymer
Acrylic acid/(C10-30) alkyl methaacrylate copolymer	Acrylic acid/(C10–30) alkyl methaacrylate copolymer
Acrylates copolymer	Aqueous solution of acrylates copolymer
Polyacrylate-1	Crosslinked copolymer of vinylpyroridone,methacrylate N, N-dimethylaminoethyl, stearyl acrylate, tripropylene glycol
Sodium acrylate copolymer	Crosslinked copolymer of sodium polyacrylate
Sodium polyacrylate	Sodium polyacrylate
Acryloyldimethyltaurate/VP copolymer	
Methylcellulose	Methylcellulose
Hydroxyethylcellulose	Hydroxyethylcellulose
Hydroxypropylcellulose	Hydroxypropylcellulose
Hydroxyethyl ethylcellulose	Hydroxyethyl ethylcellulose
Cellulose gum	Cellulose carboxymethyl ether sodium
Algin	Sodium salt of alginic acid
Carrageenan	Carrageenan
Ceratonia siliqua gum	Ceratonia siliqua gum
Hydroxypropyl starch phosphate	
Xanthan gum	

Table 2.14 Polyquateniums and constituents

INCI	Constituents
Polyquaternium-4	Hydroxyethyl cellulose dimethyl diallylammonium chloride copolymer
Polyquaternium-5	Copolymer of acrylamide and quaternized dimethylammoniumethyl methacrylate
Polyquaternium-6	Poly(diallyldimethylammonium chloride)
Polyquaternium-7	Copolymer of acrylamide and diallyldimethylammonium chloride
Polyquaternium-10	O-(2-hydroxy-3-(trimethylammonium)propoxy) hydroxyethyl chloridecellulose
Polyquaternium-11	Copolymer of vinylpyrrolidone and quaternized dimethylaminoethyl methacrylate
Polyquaternium-16	1H-Imidazolium 1-ethenyl-3-methyl-chloride polymer with 1-ethenyl-2-pyrrolidinone
Polyquaternium-22	Copolymer of acrylic acid and diallyldimethylammonium chloride
Polyquaternium-24	O-(2-hydroxy-3-(lauryldimethylammonium)propoxy) hydroxyethyl chloridecellulose
Polyquaternium-28	Copolymer of vinylpyrrolidone and methacrylamidopropyl trimethylammonium
Polyquaternium-32	Poly(acrylamide 2-methacryloxyethyltrimethyl ammonium chloride)
Polyquaternium-33	Copolymer of quaternary acrylate salt and acrylamide
Polyquaternium-37	Poly(2-methacryloxyethyltrimethylammonium chloride)
Polyquaternium-39	Terpolymer of acrylic acid, acrylamide and diallyldimethylammonium chloride

(continued)

Table 2.14 (continued)

INCI	Constituents
Polyquaternium-43	Copolymer of acrylamide, acrylamidopropyltrimonium chloride, 2-amidopropylacrylamide sulfonate, and DMAPA monomers
Polyquaternium-44	Polymeric quaternary ammonium salt consisting of vinylpyrrolidone and quaternized imidazoline monomers
Polyquaternium-46	Terpolymer of vinylcaprolactam, vinylpyrrolidone, and quaternized vinylimidazole
Polyquaternium-47	Terpolymer of acrylic acid, methacrylamidopropyl trimethyl ammonium chloride, and methyl acrylate
Polyquaternium-49	Copolymer of methacryloyl ethyl betaine, PEG-9 methacrylate andmethacryloyl ethyl trimethyl ammonium chloride
Polyquaternium-51	Copolymer of 2-methacryloyloxyethyle phosphorylcholine and butylmethacrylate
Polyquaternium-52	Copolymer of *N,N*-dimethylamino methacryl diethylsufate, *N,N*-dimethyl acrylamide dimethacryate polyglycol
Polyquaternium-53	Copolymer of acrylic acid, acrylamide and methacrylamidopropyltrimonium chloride monomers
Polyquaternium-55	Trimonium chloride polymer of vinylpyrrolidone, dimethylaminopropyl methacrylamide and methacryloylaminopropyl lauryldimoniumchloride
Polyquaternium-56	Polymeric quaternary ammonium salt consisting of isophorone diisocyanate, butylene glycol, and dihydroxyethyldimonium methosulfate monomers
Polyquaternium-57	Polymeric quaternary ammonium salt consisting of castorisostearate succinate (q.v.) and ricinoleamidopropyltrimonium chloride (q.v.) monomers
Polyquaternium-61	5% Polyalcohol solution of Lipidure®. The polyalcohol consists of glycerine and 1,3-butyleneglycol
Polyquaternium-64	Polymeric quaternary ammonium salt consisting of MPC and 2-hydroxy-3-(2-methyl-2-propenoyl)oxypropyltrimethylammonium chloride
Polyquaternium-65	5% aqueous solution of binary copolymer of ammonium chloride. Contains 1% phenoxyethanol
Polyquaternium-67	Polymer of hydroxyethyl cellulose with dimethyldodecylammonium and trimethylammonium
Polyquaternium-68	Polymeric quaternary ammonium salt formed by reacting vinylpyrrolidone, methacrylate amide, vinylimidazol, and 3-methyl-1vinylimidazolinium methyl sulfate
Polyquaternium-69	Polymeric quaternary ammonium salt composed of vinyl caprolactam, vinylpyrrolidone, dimethylaminopropyl methacrylamide (DMAPA), and methacryloylaminopropyl lauryldimonium chloride
Polyquaternium-72	Polymeric quaternary ammonium salt of hydroxethylcellulose reacted with a coco-alkyl dimethyl ammonium substituted epoxide
Polyquaternium-74	Amphoteric copolymer of acrylic acid, anionic monomer and proprietary cationic monomers
Polyquaternium-75	Natural water-soluble polysaccharide derivative produced from starch
Polyquaternium-86	Polymeric quaternary ammonium salt of vinylpyrrolidone, 1-methyl-3-vinylimidazoline chloride, vinylimidazole and methacrylic acid
Polyquaternium-87	Polymeric quaternary ammonium salt of vinylpyrrolidone, vinylimidazole and diallyldimethyl ammonium chloride
Polyquaternium-92	Polymeric quaternary ammonium salt

2.3.2.2 Improving the Texture

Polymers are also widely used for improving the texture of shampoos, conditioners, and leave-on hair care products. For example, cationic polymers are added to shampoos to reduce the "coarse" and "stuck" feeling of the hair while rinsing and improve the feel of use. Polymers used in shampoos are listed in Table 2.15.

Conditioners and leave-on hair care products contain cationic surfactants and oils to give favorable texture but may also contain polyquaternium-7, polyquaternium-10, polyquaternium-11, polyquaternium-39, and guar hydroxypropyl trimonium chloride to further improve the feel of use.

Table 2.15 Polymers used in shampoos

Polyquaternium-7	Copolymer of acrylamide and diallyldimethylammonium chloride
Polyquaternium-10	*O*-(2-hydroxy-3-(trimethylammonio) propyl) hydroxyethyl chloridecellulose
Polyquaternium-22	Copolymer of acrylic acid and diallyldimethylammonium chloride
Polyquaternium-24	*O*-(2-hydroxy-3-(lauryldimethylammonio) propyl) hydroxyethyl chloridecellulose
Polyquaternium-28	Copolymer of vinylpyrrolidone and methacrylamidopropyl trimethylammonium
Polyquaternium-39	Terpolymer of acrylic acid, acrylamide and diallyldimethylammonium chloride
Polyquaternium-47	Terpolymer of acrylic acid, methacrylamidopropyl trimethyl ammonium chloride, and methyl acrylate
Polyquaternium-52	Copolymer of *N,N*-dimethylamino methacryl diethylsufate, *N,N*-dimethyl acrylamide dimethacryate polyglycol
Polyquaternium-53	Copolymer of acrylic acid, acrylamide, and methacrylamidopropyltrimonium chloride monomers
Polyquaternium-67	Polymer of hydroxyethyl cellulose with dimethyldodecylammonium and trimethylammonium
Polyquaternium-75	Natural water-soluble polysaccharide derivative produced from starch
Trigonella foenum-graecum hydroxypropyl trimonium chloride	
Caesalpinia spinosa hydroxypropyl trimonium chloride	
Locust bean hydroxypropyl trimonium chloride	
Guar hydroxypropyl trimonium chloride	2-Hydroxy-3- (trimethylammonio) propyl ether chloride

Reference cited from Toho Chemicals

Merquat

Polymers called Merquat (Lubrizol Corporation) are available for improving the feel of use of rinse-off cosmetics such as shampoos and conditioners (Table 2.16). They are copolymers consisting of five kinds of monomers: diallyldimethylammonium, methacrylamidopropyl trimethyl ammonium chloride (which are both cationic), acrylate (anionic), acrylamide (amphoteric), and methyl acrylate (nonionic).

The feel of use of shampoos and rinse-off conditioners can be adjusted by selecting appropriate combinations and ratios of monomers (Fig. 2.3).

Table 2.16 Representative Merquat polymers

Merquat 100	Polyquaternium-6
Merquat 280	Polyquaternium-22
Merquat 295	Polyquaternium-22
Merquat 550	Polyquaternium-7
Merquat plus 3330	Polyquaternium-39
Merquat plus 3331	Polyquaternium-39
Merquat 2001	Polyquaternium-47
Merquat 2003	Polyquaternium-53

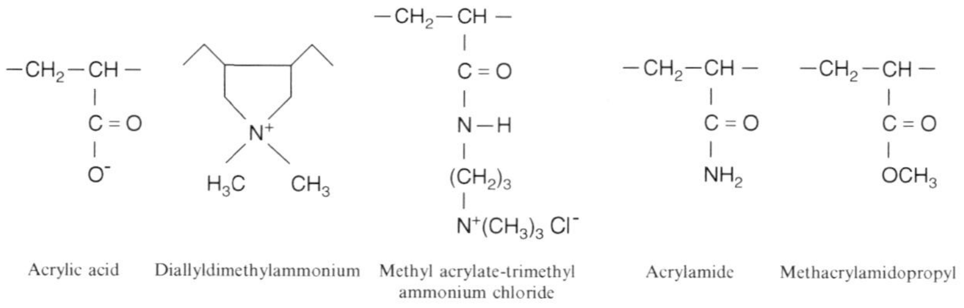

Fig. 2.3 Monomers used in Merquat

2.3.2.3 Hair Setting

There are anionic, nonionic, cationic, and amphoteric hair styling polymers, which vary in constituent depending on manufacturer. Anionic hair setting polymers consists of monomers of acrylic acid, methacrylate, and crotonic acid. Cationic polymers are copolymers of dimethylaminopropyl methacrylamides, diallyldimethylammonium chloride, propyl trimethylammonium chloride, etc. Nonionic polymers are polymers of pyrrolidone, vinyl alcohol, etc. (Table 2.17)

Hair styling polymers are added either alone or combined into hair styling and setting products so as to constitute about 10%. The setting and styling performances, spread of the hair and the feel of use vary by polymer. The feel of use given by carboxy vinyl polymers and stearyl trimethylammonium chloride is shown in Table 5.12 (Sect. 5.5, Chap. 5).

Table 2.17 Hair setting polymers

Name	Constituents		Characteristics	Structure
Acrylates copolymer	Copolymer of acrylic acid and methacrylic acid building blocks	Luvimer 100P, eo% (BASF), Plasize L-8011 (Osaka Organic Chemical Industry)	Anionic	Acryl
Ammonium acrylate copolymer	Copolymer of acrylic acid and acrylamide ethyl acrylate	Yodosol GH800F, GH34F (AkzoNobel)	Anionic	Acryl
AMP-Acrylates copolymer	Acrylic resin alkanolamine solution	Plasize L series (Goo Chemical Company)	Anionic	Acryl
TEA-Acrylates copolymer	Acrylates copolymer	Aniset (Osaka Organic Chemical Industry)	Anionic	Acryl
Acrylates/acrylamide copolymer	Acrylates/acrylamide/ethyl acrylate copolymer	Ultrahold, Strong	Anionic	Acryl
Acrylates/diacetoneacrylamide copolymer	Acrylic resin alkanolamine solution	Plascize L series (Goo Chemical Company)	Anionic	Acryl
Dimethyl acrylamide/hydroxyethyl acrylate/methoxyethyl acrylate copolymer		Plascize L-2700 (Goo Chemical Company)	Anionic	
AMPD-Acrylates/diacetoneacrylamide copolymer		L-1260, L-53D (Goo Chemical Company)	Anionic	
AMP-Acrylates (C1-C18)/diacetoneacrylamide (C1-8) copolymer		Plascize L series (Goo Chemical Company)	Anionic	
Acrylates/octylacrylamide copolymer	Acrylic octylamide/acrylate ester copolymer	AMPHOMER V-42 (AkzoNobel)	Anionic, neutralization required	
Ethyl ester of PVM/MA copolymer	Monoalkyl ester copolymer of vinyl methyl ether and maleic anhydride	Gantrex ES-225, ES335 (ISP)	Anionic, neutralization required	
Polyurethane-14 AMP-acrylates copolymer	Polyurethane-n and acrylates copolymer	DynamX (AkzoNobel)	Anionic	Polyurethane/acryl
PEG/PPT/dimethicone/acrylates copolymer		Luviflex Silk (BASF)	Anionic	
VA/crotonates/vinyl neodecanoate copolymer	Copolymer of vinyl acetate, crotonates and neodecanoates	Resyn 28-2930 (AkzoNobel)	Anionic	Vinyl acetate

(continued)

Table 2.17 (continued)

Name	Constituents		Characteristics	Structure
AMP-Acrylates/hydroxypropyl acrylate/dimethylaminoethylmethacrylate/diacetone acrylamide/VP copolymer		Naturesin/AP Polymer 560 (Osaka Organic Chemical Industry)	Amphoteric	
Octylacrylamide/hydroxypropyl acrylate/buylaminoethylmethacrylate copolymer	Copolymer of octylacrylamide, hydroxypropyl acrylate and buylaminoethylmethacrylate	AMPHOMER 28-4910, SH30, SH37, SH701 (AkzoNobel)	Amphoteric, hard setting	
Polyquaternium-49	Methacryloyl ethyl dimethyl betaine/methacrylate ethyl trimethyl ammonium chlorido/methosy polyethylene glycol methacrylate copolymer solution	WetResin (Osaka Organic Chemical Industry)	Betaine system, plastic, water holding	
Polymethacryloyl ethyl betaine	Ethyl betaine polymer of crylates	Plascize L-401 (Goo Chemical Company)	Amphoteric, hair conditioner	Acryl
Methacryloyl ethyl dimethyl betaine/methacrylate ethyl ammonium chlorido/hydroxyethyl methacrylate copolymer	Polymer consisting of cationic methacrylates and copolymers of methacryloyl ethyl dimethyl betaine and hydroxyethyl methacrylate	Plascize L-450 (Goo Chemical Company)	Amphoteric & cationic (betaine system), hair conditioner	
Methacryloyl oxyethyl carboxy betaine/acrylate copolymer		REM Resin (Osaka Organic Chemical Industry)	Amphoteric, hard setting, neutralization required	
Methacryloyl ethylbetaine/acrylate copolymer	*N*-Methacryloyl oxyethyl *N,N*-dimethyl ammonium-α-*N*-methyl carboxy betaine alkyl methacrylate copolymer	Yukaformer R205, R205S, 510, 301 (Mitsubishi Chemical)		
Isobutylene/ethylmaleimide/hydroxyethylmaleimide copolymer	Imide copolymer of isobutylene and maleic anhydride	Aquaflex EX-64 (ISP)		
VP/vinyl caprolactam/DMAPA acrylates copolymer	Terpolymer of vinyl caprolactam, vinylpyrrolidone and dimethylaminopropyl methacrylamide	Aquaflex SF-40 (ISP)		Physiolocially inactive to human, safe
Vinyl caprolactam/VP/diethylaminoethyl methacrylamide copolymer	Terpolymer of vinyl caprolactam, vinylpyrrolidone and dimethy laminoethyl methacrylamide	Copolymer VC-731 (ISP)	Cationic	

VP/DMPA acrylate copolymer	Copolymer of vinylpyrrolidone and dimethylaminopropyl methacrylamide	Styleze CC-10 (ISP)	Cationic	
Polyquaternium-11	Copolymer of vinylpyrrolidone and quaternized dimethylaminoethyl methacrylate	Gafquat 734, 755N (ISP), Luviquat PQ11PN (BASF)	Cationic	Highly compatible with anionic, cationic and nonionic ingredients
Polyquaternium-16	3-Methyl-1-vinylimidazolium chloride-1-vinyl-2-pyrrolidinone copolymer	Luviquat FC370, 550, 905 (BASF)	Cationic	
Polyquaternium-28	Copolymer of vinylpyrrolidone and methacrylamidopropyl trimethylammonium chloride	Gafquat HS-100 (ISP)	Cationic	
Polyquaternium-44	Vinylpyrrolidone and quaternized imidazoline monomers	Luviquat Care (BASF)		
Polyquaternium-55	Trimonium chloride polymer of vinylpyrrolidone, dimethylaminopropyl methacrylamide and methacryloylaminopropyl lauryldimoniumchloride	Styleze W-10 (ISP)	Polymeric surfactant, wetting agent	
VP/VA copolymer	Copolymer of vinyl acetate and vinylpyrrolidone		Nonionic	Soluble to organic solvents. Amphipathic copolymer
PVP	Polyvinylpyrrolidone		Nonionic	Vinyl acetate
VA/acrylates/lauryl methacrylate copolymer		Styleze 2000 (ISP)	Nonionic	
VA/methacrylamide/vinyl imidazole copolymer		Luviset Clear (BASF)	Nonionic	
Polyvinylcaprolactan		Luviskol Plus (BASF)	Nonionic	

2.3.2.4 Stabilizing Emulsions

Polymers are also used as stabilizers of creams and milky emulsions. Xanthan gum, carboxyvinyl polymer, and crosspolymers of acrylate and C10–30 alkyl methacrylate are anionic polymers used for stabilizers. Cationic polymer of guar hydroxypropyl trimonium chloride is also appropriate for the use. The polymers are combined so as to constitute 0.5% or less. Polymers hardly undergo changes in viscosity by temperature changes and stabilize emulsions, but may coagulate in emulsions and destabilize them; and therefore they should be thoroughly tested.

2.3.3 Dissolving Polymers

Powder polymers can be dissolved in solvents, such as water, by directly adding into the solvent and stirring long at high speed by using a disperser. Polymers can be dissolved easier by increasing the surface area of the powder by dispersing in liquid into which the polymer does not dissolve or mixing with another kind of powder. Techniques for dissolving polymers are described below.

2.3.3.1 High-Speed Stirring

Add the polymer gradually while stirring the solvent at high speed by using a disperser. This method is appropriate for Ulterz 10, PEMULEN TR-1, and TR-2, which are carboxyvinyl polymers.

2.3.3.2 Changing the pH

Polymers that easily dissolve at the neutral pH range can be dissolved easily by dispersing in low pH water, which is prepared by adding acid, and then increasing the pH by adding alkali. This method dissolves carboxyvinyl polymers such as Carbopol 980 and 981 easily even without the use of a disperser.

2.3.3.3 Dispersing in Glycol

To dissolve polymers that dissolve in water but are prone to forming lumps and require long-time stirring, it is recommended to disperse the polymer in glycol and then add water and dissolve. Most cellulose polymers do not dissolve in glycols. Therefore, powder polymers, such as xanthan gum and cellulose carboxymethyl ether sodium, can be dissolved easily by dispersing into glycerin or 1,3-butylene glycol and adding water to the dispersion.

2.3.3.4 Dispersing in Cold Water and Heating

Polymers that are highly water soluble at high temperatures but are little soluble to cold water, such as hydroxyethylcellulose and polyquaternium-10, can be easily dissolved by dispersing in cold water and then heating.

2.3.3.5 Mixing with Other Powder

It is also effective to mix powder polymer with other water-soluble powder such as salt before adding into water. Dispersion of the polymer into the other powder prevents it from forming lumps. In such a case, the quantity of the powder in the product must be larger than that of the polymer. Water-soluble powder such as salt, mannitol, and sorbitol can be used.

2.3.3.6 Dispersing in Liquid Oil

Most polymers are insoluble to oils. For preparing products that contain liquid oil, such as squalane, liquid paraffin, liquid ester, and vegetable oil, polymers can be dissolved by dispersing into a portion of the liquid oil and then adding the dispersion into water. The preparation involves dispersing the polymer in the liquid oil, adding the dispersion into the aqueous phase, dissolving it by stirring while heating to the predetermined emulsification temperature, adding the oil phase and emulsifying. The time needed for dissolving the polymer is the time until the water reaches the emulsification temperature, and the process does not require additional time for dissolving the polymer. While the polymer is dissolved, the mixture separates into water and oil, but the polymer gradually dissolves into water resulting in increased surface area and accelerated dissolution of the polymer. Instead of dispersers and homogenizing mixers, low-speed paddles can be used. This method can easily and quickly dissolve carboxyvinyl polymers and xanthan gum.

2.4 Glycols

Glycols are alcohols but have higher melting points and viscosity than ordinary alcohols. They freely dissolve to water, and the polarities, which are determined by the OH group and the number of carbons, are intermediate of surfactants. They can be used for various purposes by using the properties and are very useful in preparing cosmetics.

2.4.1 *Kinds of Glycols*

Widely used glycols include glycerin, 1,3-butylene glycol, propylene glycol, 3-methyl-1,3-butyldiole, polyethylene glycol, diglycerin, and dipropylene glycol (Table 2.18).

2.4.2 *Purposes of Using Glycols*

2.4.2.1 Moisturizer

Glycols are combined into creams, facial toners and leave-on hair care cosmetics as moisturizers. They have direct effect on the feel of use of the leave-on products.

2.4.2.2 Stabilizer

Glycols are added as an emulsion stabilizer so as to constitute several percent into formulations mainly consisting of a combination of water, surfactants and thickening polymers, to solubilize the oily components.

2.4.2.3 Solvent and Resolvent

Glycols are used as solvents of polar substances insoluble to oils, such as paraben.

They are also used to solubilize polar oils insoluble to water to make transparent products. They can also reduce the amount of ethanol in the formulation.

Isoprene glycol, 1,3-butylene glycol and propylene glycol are effective for adding parabens, menthol, sebacate, and adipate into transparent formulations.

2.4.2.4 Solubilizer of Aroma Chemicals

When aroma chemicals are solubilized into liquid formulations, the nonionic surfactant components such as polyoxyethylene alkylether surfactants and polyoxyethylene hydrogenated castol oil may undergo hydration and gelling. The aroma chemicals can be easily solubilized by adding them into a mixture of 1,3-butylene glycol, propylene glycol, and nonionic surfactants.

2.4.2.5 Disperser of Polymers

Dissolving polymers in water may require long stirring by using dispersers because polymers are prone to forming lumps. They can be easily dissolved by dispersing the polymers in glycerin and then adding the dispersion into water.

Table 2.18 Feels given by glycols

8% solution		Glycerin	1,3-Butyleneglycol	Propreneglycol	Isoprene glycol	Diglycerin	Dipropylene glocol
Feel	Skin	Rather moisturized	Moisturized	Very smooth	Rather smooth	Smooth and moisturized	Soft, smooth
	Hair	Sticky	Rather sticky	Very smooth and fresh	Fresh and light	Rather sticky	Fresh

Summary: Polarity: Isoprene glycol < 1,3-BG < PG < Glycerin; Hydrophilicity: Isoprene glycol < 1,3-BG < PG < Glycerin
The higher the polarity, the higher the hydrophilicity. High hydrophilicity results in high moisturization and give moisturized feels. Diglycerin and propreneglycol are dimers and give heavy feelings

2.5 Other Ingredients

2.5.1 Ingredients for Giving Efficacies, Effects, and Concepts

Each cosmetics manufacturer uses original combinations of ingredients that give efficacies, effects, and concepts. Manufacturers also use original components that can attract users. Up to present, about 1,000 ingredients have been proposed and used, which can be classified as described below.

2.5.1.1 Plant Extracts and Herbal Medicine Components

There are approximately 800 ingredients of plant origin, which are used extracted in water, 1,3-butylglycol, ethanol, etc. or in the form of powder. Plant extracts are combined mainly for the concepts of whitening, antiaging, wrinkle prevention, anti-inflammatory, acne prevention, moisturizing, slimming, scalp care, hair growth, and damage hair treatment.

2.5.1.2 Microbial-Derived Ingredients

Ingredients derived from microbes include Saccharomyces ferment filtrates and Lactobacillus ferment filtrates. They are used in facial toners and creams for the concepts of wrinkle prevention and antiaging.

2.5.1.3 Proteins and Amino Acids [3]

Proteins that are used in cosmetics include collagen, keratin, and proteins extracted from soy beans, silk, milk, wheat, pearl, sesame, etc. To be combined in cosmetics, the proteins are hydrolyzed into smaller molecules or modified into derivatives, such as quaternized ammonium and acylated, siliconized and esterified derivatives [2].

2.5.1.4 Ceramides

Ceramides are lipids of biological origin and include Ceramide 1, Ceramide 2, Ceramide 3, Ceramide 6II, phytosteryl isostearate, phytosphingosine, cholesterol, and phytosterol.

2.5.1.5 Vitamins

Vitamins, such as ascorbic acid phosphate magnesium salts, pyridoxine, retinol palmitate, and tocopherol, are useful in cosmetics.

2.5.2 Coloring Agents

Coloring agents for cosmetics include anionic dyes and natural colorants.

Certified anionic dyes have a sulfonic or carboxylic group. The coloring group (such as nitro, nitroso, phenol, triphenylmethane, anthraquinone, and quinoline) bound to the benzene framework, etc. of the dye determine the color it develops. They are soluble to water and ethanol. Many are somewhat stable to heat, light, and low and high pHs; but few dyes resist oxidization and reducing agents; and many fade by ultraviolet rays, solar light, and fluorescent light.

Most common natural colorants are plant pigments, such as gardenia yellow, carotene, turmeric, paprika, safflower, *Lithospernum* purple, sodium copper chlorophyllin, and sodium guaiazulene sulfonate. There are also animal pigments including cochineal and laccaic acid. The colors change by pH are relatively low resistant against light and heat, and are vulnerable to fading and color changes. Colorants produced from plant extracts include *Glycyrrhiza glabra* root extract and *Phellodendron* extract and are mainly yellow or brown. Caramel is a coloring agent produced from saccharides (Table 2.19).

2.5.3 Scenting Agents

Widely used scenting agents include essential oils of orange, grapefruits, eucalyptus, rosemary, mint, geranium, lavender, and Damask rose (Rosa × damascena). Perfume components include phenethyl alcohol, geraniol, citronellal, menthol, and linalool.

2.5.4 Ingredients for Stabilizing Products

There are ingredients for stabilizing and maintaining the quality of cosmetics products.

They are indispensable for most products because cosmetics are used over a long period after opening the package and may be vulnerable to deterioration and putrefaction. Because they are applied on delicate skin, they must be highly stable. Ingredients for stabilizing the products are classified into preservatives, pH adjusters, antioxidants, and chelating agents (Table 2.20).

2.5.4.1 Preservative-Free Cosmetics

Many preservative-free cosmetics have been put on the market. Their concept is not including specific ingredients such as preservatives, silicones, sodium laureth sulfate and sodium lauryl sulfate (for shampoos).

Table 2.19 Natural pigments and characteristics [2]

	Carthamin	Safflower yellow	Guaiazulene	Shikonin	Sodium copper-chlorophyllin	Laccaic acid
Primordia	Extracted from *Carthamus tinctorius* flowers in slightly alkaline water	Extracted from the tubular flower stage of *Carthamus tinctorius*	Guaiol from *Lignum vitae* essential oil	Crystals extracted from *Lithospermum erythrorhizon*	Porphyrin	Prepared from shellac secrete from female *Laccifer lacca*
Principal ingredients	Flavonoid carthamin		Dimethyliso-propylazulene		Salt of sodium and copper chlorophylls a and b	Natural anthraquinone dicarboxylic acid
Chemical formula/molecular weight	Carthamin $C_{21}H_{32}O_{11}$:236.3	$C_{21}H_{22}O_{11}$:450.4	Guaiazulene $C_{15}H_{18}$:198.3	Alkhanin $C_{16}H_{19}O_5$:288.3	$C_{34}H_{30}O_5N_4$ $CuNa_2$:684. $2C_{34}H_{28}O_6N_4CuNa_2$: 684.2	Laccaic acid A $C_{26}H_{19}NO_{12}$:537.4, B $C_{24}H_{16}NO_{12}$:496.4
Color	Dark red to dark purple	Yellow	Blue	Red	Blackish green to bluish black	Orange red
Solubility	Insoluble to water	Dissolves in cold water	Insoluble to water	Dissolves in organic solvents		Hardly soluble to water
		Easily soluble to thin ethanol and thin PG and forms clear solutions	Dissolves in organic solvents	Dissolves in alcohol		Dissolves in alkaline
pH		Insoluble in acid	Color changes by pH			Color changes by pH
	Fades by alkaline		Red in acid and bluish violet, blue or purple in alkaline			Orange to orange red in acid, red in neutral and purplish red in slight alkaline
Light resistance	Weak	Stable at neutral pHs			Stable	Stable in acid
Heat resistance	Stable	Stable at 100°C for 1 h			Good	Stable in acid, less stable in alkaline

Metal ion		Darkens in presence of Fe				Deteriorates by presence of metal ions
Oxygen			Anti-inflammatory, antibacterial and fungicidal action	Antimicrobial activity		
Safety		Acute toxicity LD50: 20 g/kg (oral)	Acute toxicity LD50: 1.22 mg/kg (oral)			Acute toxicity LD50: 3,300 mg/kg (oral)
	Turmeric extract	β-Carotene	Gardenia blue	Gardenia yellow	Cochineal	Paprika
Primordia	Diketone system pigment	Carotenoid pigment	Extracted in ethanol from the fruit of gardenia	*Rubiaceae gardenia*, iridaceae saffron crocus, mahogany	Prepared from dry powder of female cochineal insect	Extracted in hexane from a red pepper
		Widely synthesized				
Principal ingredients	Curcumin	One of provitamins		A kind of carotenoid	Kind of the anthraquinone derivative	Carotenoid
Chemical formula/ molecular weight	Curcumin $C_{12}H_{20}O_6$:368.4	Carotene $C_{30}H_{56}$:536.9			Carminic acid $C_{22}H_{20}O_{13}$:492.4	Capsaicin $C_{30}H_{59}O_3$:597.0, capsorubin $C_{30}H_{60}O_4$:605.0
Color	Yellow-dark orange	Purplish to dark red	Dark blue to dark livid	Yellow-orange	Red to dark reddish brown	Orange to reddish brown
Solubility	Extracted into BG-water and PG-water	Insoluble to water, PG and ethanol	Insoluble to ethanol	Soluble to water, thin ethanol and thin PG	Soluble to water, ethanol and PG	Insoluble to water
		Little soluble in organic solvents	Dissolves and forms clear solutions in 50% PG and water			Easily soluble to acetone and chloroform
pH	Lemon yellow in acid	Acidic conditions promote decomposition.	Forms coagulation settling in acid	Decomposes by acid	Changes by pH (easily dissolvable in sodium hydroxide solution)	Most stable at pH 6

(continued)

Table 2.19 (continued)

	Carthamin	Safflower yellow	Guaiazulene	Shikonin	Sodium copper-chlorophyllin	Laccaic acid
	Darkly reddish brown in alkaline	Relatively stable in weak alkaline		Stable at pH 7–8	Yellow at pH 3, red at pH 4–5 and purplish red at pH 6–8	
Light resistance	Turns brown	Instable		Faces by acid		Weak to acid and most stable at pH6
Heat resistance	Relatively strong	Disintegrates		Decomposes by acidity		Relatively stable (at 160°C, 2 h)
Metal ion				Deteriorates in presence of metal ions	Forms water-insoluble salts in presence of metal ions	Fades in presence of Fe, Cu or Co ions
Oxygen				Oxidizes		
Safety	Daily intake capacity (WHO): 0–2.5 mg/kg		Acute toxicity LD50: 16.7 g/kg (oral)			Acute toxicity LD50: 24 g/kg (oral)

Table 2.20 Ingredients for stabilizing products

Antimicrobial agents	Paraben, benzoate, sorbic acid, methylisothiazolinone
pH control chemicals	Salts: sodium citrate, sodium lactate, sodium hydrogen phosphate, etc.
	Acids: citric acid, phosphoric acid, glycolic acid, lactic acid, etc.
	Alkalis: NaOH, KOH, TEA, AMP, etc.
Antioxidants	Tocopherol, BHT, lecithin, sodium sulfate
Chelating agents	EDTA, hydroxyethane, diphosphonic acid

The ingredients have been indicated to somewhat affect the safety, and the inclusion had to be stated under the former Pharmaceutical Affairs Law. Preservatives, sodium laureth sulfate and sodium lauryl sulfate, have a risk of irritating the skin. Silicone, which is highly hydrophobic, may form a film on the skin, entirely blocking water and thus adversely affecting the health of the skin. On hair, it obstructs permanent waving and coloring.

On the other hand, the ingredients have many benefits and are still used in many cosmetics. For example, paraben is the most widely used preservative in the USA and Canada [5], and silicone is an efficacious ingredient in skincare, makeup, and leave-on hair care products. They are also used in products other than cosmetics and are stable and safe. Sodium laureth sulfate and sodium lauryl sulfate are widely used in shampoos and toothpastes as detergents, emulsifiers, and lathering agents.

The concept of preservative-free is used mainly for marketing purposes rather than for ensuring safety. Formulations of cosmetics should be designed by comprehensively and objectively judging functions, stability, costs, etc. as well as safety (Table 2.21).

Table 2.21 Preservatives and germicides used in preservative-free formulations [4, 5]

Name	Amount	Notes
1,2-Hexanediol		Moisturizer. May lower the viscosity of the creams
Pentylene glycol		Moisturizer
Caprylyl glycol		Moisturizer. May solidify at low temperatures
Octoxy glycerin or ethylhexyl glycerin	ca. 0.2%	Multifunctional preservative
Glyceryl caprylate		Moisturizer
Ethylhexyl glycerin/glyceryl caprylate		Mixture of nonionic surfactants. Can be also combined into nonmedicinal products
Polyglyceryl-2 laurate		Mixture of nonionic surfactants. Emulsifier of a wide antibacterial spectrum
Sodium coco PG-dimonium chloride phosphate		Highly safe. Not suitable for Carbopol, sodium acrylate and CMC
Alkyl diaminoethylglycine	0.03–0.3%	Amino acid type amphoteric surfactant
PCA ethyl cocoyl arginate		Cationic surfactant
Carpyloyl glycine		Dissolved when neutralized with Na or K hydroxide. Precipitates at low pH
Phenoxyethanol	1% or less	Effective when combined with cationic surfactants

(continued)

Table 2.21 (continued)

Name	Amount	Notes
Citrus grandis seed extract	ca. 0.5%	Plant extract
Cinnamon extract, potassium sorbate, ethanol[a]		Little soluble to water
Clove extract, *Artemisia capillaris* extract, glyceryl caprylate, polyglyceryl-10 laurate, BG[b]	Around 1%	Adjust pH at 6 or lower with organic acid
Clove extract, Artemisia capillaris extract, glyceryl caprylate, BG[c]	Around 0.5%	Adjust pH at 6 or lower with organic acid

[a]Natrulon PC-15
[b]SY Plantex KTB (Sakamoto Yakuhin) [6]
[c]SY Plantex KN (Sakamoto Yakuhin) [6]

References

1. L. DiBerardino (1999) *Cosmetic & Toiletries magazine*, Allured Publish Corp., 112(10)
2. H. Tamura et al. (2006) New Cosmetics Handbook, Nikko Chemicals Co., Ltd., 132 (in Japanese)
3. M. Yoshioka, Promoise-peptide renaissance, Seiwa Kasei Co., Ltd. (in Japanese)
4. R. L. Chapman (2006) *Cosmetics & Toiletries magazine*, Allured Publish Corp., 125(11)
5. T. Yamada, S. Oshima (2006) Fragrance Journal, Fragrance Journal Ltd., 34(4) (in Japanese)
6. Fujimito Takehiko (1981) Introduction to surfactants—New Edition (in Japanese), Sanyo Chemicals Industries

Chapter 3
Emulsions

Abstract Emulsions consist of two liquid phases that are mutually insoluble and have one liquid (i.e. the dispersed phase) being dispersed in the other (the continuous phase) in form of microparticles. Stabilization is the main target in formulating emulsions because all emulsions move toward separation. Selection of oils and emulsifiers determines the product form and involves deciding the type and ionicity of the emulsion based on the performance, appearance, and viscosity of the constituents to be included in the cosmetics. Emulsifiers bind the phases and keep them in the emulsified state and are either a surfactant or a combination of surfactants.

When an emulsion loses stability, it undergoes changes in properties such as separation, coagulation, sedimentation, viscosity change, and loss of uniformity. The major cause is inappropriate formulation, but temperature and light may also destabilize emulsions.

Keywords Continuous phase • Dispersed phase • Stability • Viscosity

3.1 Designing the Formulations of Emulsions

Emulsions, which are produced from oils and surfactants, are particularly widely used in cosmetics. Emulsification is an effective mean for combining efficacious oily and aqueous components into a single system and thus for producing efficacious cosmetics.

The properties, functions, and feel of use of emulsions are determined by the polarity, melting point, and chemical structure of the oily component and the combination and structures of hydrophilic and alkyl groups of the surfactant. It is also important to design stable systems by combining water, surfactants, and oily components.

H. Iwata and K. Shimada, *Formulas, Ingredients and Production of Cosmetics: Technology of Skin- and Hair-Care Products in Japan*, DOI 10.1007/978-4-431-54061-8_3,

Many high-performance emulsions are available on the market, but knowledge and techniques for designing emulsions of aimed properties, which involves combining materials, are indispensable for designing new cosmetics products.

Emulsions consist of two liquid phases that are mutually insoluble. One of the phases (i.e. the dispersed phase) is dispersed in the other (continuous phase) in form of microparticles. Stabilization is the main target in formulating emulsions because all emulsions move toward separation. Emulsifiers bind the phases and keep them in the emulsified state and are either a surfactant or a combination of surfactants.

Emulsions need to have the viscosity and texture appropriate for:

1. Combining the effective and useful conceptual constituents
2. Type of emulsion (either O/W or W/O) and ionicity
3. Giving the feel of use and texture appropriate for the purpose and
4. Giving the product form and viscosity that suit the container

This chapter describes the basis of selecting oils and emulsifiers, which determine the product form. It involves deciding the type and ionicity of the emulsion based on the performance, appearance, and viscosity of the constituents to be included in the cosmetics. Methods for solving problems in producing emulsions are also explained as well as those for investigating and deciding the type of emulsion, oils, and surfactants.

3.1.1 Formulation of Emulsifiers

3.1.1.1 Factors Determining the Type of Emulsion

The type of emulsion (whether it is oil-in-water (O/W) or water-in-oil (W/O)) determines the feel of use and is mainly decided based on the purpose of using the cosmetics.

O/W emulsions give nonsticky, light, and moisturized feelings. They can be used for both rinsed-off and leave-on cosmetics. W/O emulsions give an oily feeling and are used for rinsed-off and leave-on products of special performances, such as in some cleansing creams and cream foundations.

(a) Volume ratio between the dispersed and continuous phases
The phase of the larger volume (content) becomes the outer (continuous) phase.
In O/W, oil content is 1–25%.

(b) HLB (hydrophile–lipophile balance) range of each emulsion type
O/W HLB: 8 ~ 15
W/O 3 ~ 6

3.1.2 *Selecting Surfactants*

3.1.2.1 Properties of Surfactants

Surfactant to be used as emulsifier should be selected based on the body part on which the product will be used (i.e. for the skin or hair) and the method and purpose of use.

Cosmetics to be applied on the skin mainly use nonionic surfactant combined with anionic surfactant or fatty acids. On the other hand, cationic surfactant is widely used for hair cosmetics. Emulsions for leave-on hair cosmetics (such as hair waxes and hair creams) have a composition similar to that for skin cosmetics.

3.1.2.2 Combinations of Surfactants

Combination of Nonionic Surfactants as Emulsifier and Quantity

As described in Chap. 2, appropriate nonionic surfactants have a lipophilic group of either lauryl, cetyl, stearyl, or oleyl, and a hydrophilic group of either polyoxyethylene or glycerin, and include polyoxyethelene fatty acid esters, either alone or combined with fatty acids, polyoxyethylene alkyl ethers, polyglycerin fatty acid esters, and sucrose fatty acid esters.

Small molar number of the hydrophilic group, such as polyoxyethylene and glycerin, results in low HLB, hydrophobicity, and increased solubility in oil. Large molar number results in high HLB, hydrophilicity, and increased solubility in water. Two or more nonionic surfactants of different HLB values are used for preparing emulsions.

Simple combinations are easily affected by temperature and are unstable. An appropriate combination is, for example, mixing three to six portions of a nonionic surfactant of HLB 16–20, one portion of a nonionic surfactant of HLB 9–15, and two to six portions of a nonionic surfactant of HLB 8 or smaller. The resultant emulsifier can emulsify liquid paraffin, ester and higher alcohol into an O/W emulsion because the surfactant-oil ratio becomes 1:4–1:6. The optimum HLB of an emulsifier varies depending on oily constituents and is higher for oils of higher polarity and lower for oils of lower polarity (Table 3.1).

Combination of Ionic Surfactants and Quantity

Fatty Acids for Soap Emulsion

Stearic acid is the most commonly used fatty acid for this purpose. The viscosity of the emulsion is determined by the quantities of stearic acid and alkali agent.

Table 3.1 Combination of surfactants and kinds of cosmetics

Combination of surfactants	Cosmetics to be applied on the skin	Hair cosmetics	
		Rinse-off	Leave-on
Anionic and nonionic surfactants	○ (Skin creams, etc.)		○ (Hair creams)
Fatty acid, nonionic surfactants, and alkali agent	○ (Skin creams, etc.)		○ (Hair waxes, and hair creams)
Cationic and nonionic surfactants		○	○
Cationic surfactant		○	○
Nonionic surfactant	○		○

Blank indicates that it is not used and ○ indicates that there is combination of surfactants use for kinds cosmetics

When the molar numbers of alkali agent and stearic acid are similar, the resultant emulsion becomes little viscous and has a high pH. When the amount of the alkali agent is 10–20% of the molar number of stearic acid, the resultant emulsion is viscous and has a pH of about 7. This is likely due to the carboxyl group in non-neutralized stearic acid, which functions as an acid. An appropriate amount of stearic acid is 2–8% of the entire emulsion; and the larger the amount, the more viscous the emulsion becomes. Non-neutralized stearic acid also serves as a thickener. The palmitic acid content in stearic acid affects the viscosity of emulsion. When the palmitic acid content is low compared to stearic acid, the resultant emulsion becomes hard and viscous; and high stearic acid content results in a relatively nonviscous smooth emulsion. This is likely attributable to the number of carbons differing by 2 between the acids. The stearic acid salts produced are liquids at high temperatures and solids at low temperatures. The Krafft point of sodium stearate is about 50–65°C. At temperatures lower than the Krafft point, sodium stearate is not an effective emulsifier. Therefore, emulsions emulsified only by stearic acid and alkali agent undergo large changes in viscosity by temperature and are unstable at high temperatures. For soap emulsification, fatty acids need to be combined with a nonionic surfactant(s) and polymers. Nonionic surfactants serve as an emulsifier at low temperatures, and polymers undergo little viscosity changes by temperature, suppress the movements of emulsion particles and thus stabilize the emulsion at high temperatures. Alkali agents affect the properties of emulsions. Sodium hydroxide produces relatively hard creams of coarse appearance. Potassium hydroxide makes relatively soft cream of intermediate viscosity and fine and lustrous appearance. With triethanolamine, soft and low viscous cream of fine and lustrous texture is produced.

Anionic Surfactants

The same applies for using an anionic surfactant as an emulsifier. The amount of anionic surfactants does not need to be as large as that of stearic acid, and several percent of the total amount of emulsion is sufficient in general, although the amount may differ by the aimed viscosity. It is obviously necessary to combine with nonionic surfactants.

Cationic Surfactants

In most cases, emulsification by a cationic surfactant involves emulsification of the cationic surfactant alone. Cationic surfactants alone produce stable emulsions. Because they are commonly used in hair cosmetics, the feel of use is a most important target. A large amount of higher alcohols, such as cetanol, is added to the oily constituent so as to account for at least 50% in most hair cosmetics products. Cetanol, which has an OH group at an end, is highly polar, is affinitive to water, and increases the viscosity of the emulsion. Adding a cationic surfactant of about 20% of the oily constituent is sufficient for producing a stable emulsion. The viscosity of the emulsion is almost determined by the ratio between the cationic surfactant and higher alcohol. It will be described in detail in the section of designing conditioners.

3.1.3 Selecting Oily Constituents

3.1.3.1 Selecting Oily Constituents from the Viewpoint of Viscosity

The viscosity of a product (emulsion) must be adjusted so as to suit its container. Factors that determine the viscosity of a cream are the structure of the alkyl group, polarity, and melting point of oily constituents.

The structure of the alkyl group (such as whether it is linear or has side chains and whether it has double bonds or not) affects the viscosity. Alkyl groups with double bonds and side chains result in low viscosity.

Polarity also affects the viscosity. The majority of oily constituents used in cosmetics are either hydrocarbons, waxes, triglycerides, esters, higher alcohols, or fatty acids. All except hydrocarbons have oxygen atoms. Higher alcohols and fatty acids that have an OH or COOH terminal are highly polar and produce viscous emulsions. On the other hand, less polar hydrocarbons produce less viscous emulsions.

Waxes rarely become viscous because they contain large amounts of fatty acids of large molecular weights and higher alcohol esters. Triglyceride is highly polar but makes less viscous emulsions compared to higher alcohols likely because the three ester bonds do not face the outside of the molecule due to the three alkyl groups. Esters have a polarity intermediate of those of hydrocarbons and higher alcohols; and the viscosity is little affected by polarity because the oxygen atom is in the middle of the molecule. Less viscous creams can be produced by increasing the amount of hydrocarbons and reducing higher alcohols. To produce viscous creams, hydrocarbons are increased, and higher alcohols are reduced.

Melting point also affects the viscosity. Use of liquid paraffin instead of microcrystalline wax causes only a little change in viscosity because both consist of hydrocarbons of high melting points. The melting points of low-polar oily constituent have little effect on viscosity, and those of polar higher alcohols and fatty acids have great effects (Table 3.2).

Table 3.2 Factors determining the viscosity

Factors determining the viscosity of emulsion	Viscosity
Alkyl group structures of oily constituent(s) and surfactant(s)	Linear saturated chain (viscous) > Branched unsaturated (less viscous)
Polarity of oily constituent(s)	Polar (viscous) > Nonpolar (less viscous)
Melting point of oily constituents(s)	High melting point (viscous) > Low melting point (less viscous)

Relationship Between Alkyl Group Structure and Viscosity

Typical straight-chain alkyl groups of oils are lauryl (C12), myristyl (C14), palmityl (C16), stearyl (C18), arachidyl(C20) and behenyl (C22) and oleyl, linole and linolen, which have one, two, and three double bonds, respectively. Linole and linolen are easily oxidized and are thus rarely used for cosmetics. Branched chains include hexyldodecyl (C16), isostearyl (C18) and octyldodecyl (C20).

The relationship between the alkyl group and viscosity was investigated by preparing emulsions using the same formulation but different oils. The oils tested were:

1. Higher alcohols (C12 ~ C22) used as basic components of creams
2. Oleyl alcohol, which has a double bond and produces soft cream and milky lotion by itself and
3. Oils that have a branched chain (hexyldodecanol (C16), isostearyl alcohol (C18) and octyldodecanol (C20))

From the investigation, the following conclusions were obtained:

1. The relationship between the length of the alkyl group and viscosity was particularly apparent in higher alcohols. With the larger number of carbons, the more viscous emulsion was produced. Higher melting point was also predicted to result in higher viscosity. Actually, the viscosity was the lowest when lauryl alcohol (C12) was used, followed by myristyl alcohol (C14), and palmityl alcohol (C16) as predicted; but stearyl alcohol (C18), arachidyl alcohol (C20) and behenyl alcohol (C22) resulted in lower viscosity compared to palmityl alcohol. This was likely because the effect of the OH group at the end of the chain, which gives hydrophilicity to the higher alcohols, is low in the alcohols that have more than 16 carbon atoms and thus they are rather hydrophobic. The most hydrophilic alcohol is likely to have 16 carbons.
2. Oleyl alcohol is easy to produce soft creams, and an emulsion consisting of oleyl alcohol alone can produce fluid creams and milky lotions. Both *cis* and *trans* double bonds are available. It is slightly more difficult to obtain uniform orientation compared to straight-chain alcohols.
3. Hexyldodecanol (C16), isostearyl alcohol (C18), and octyldodecanol (C20), which have a branched chain structure, tend to result in low viscosity. The

branched chain of the alcohols causes steric hindrance, which disturbs uniform orientation of the molecules and thickening.

Of the higher alcohols that have 18 carbons, straight-chained stearyl alcohol produces stable and viscous creams, which undergo increases in viscosity as time goes by. Oleyl alcohol produces soft creams. Octyldodecanol becomes low viscous creams and cannot act as an emulsifier by itself. Straight alkyl chains result in high viscosity, oils with a double bond tend to show low viscosity, and those with a branched chain show even lower viscosity.

3.1.3.2 Selecting Oily Constituents from the Viewpoint of Stability

Melting Point and Polarity

Constituents should be selected so as to accomplish the purposes. In principle, oils are decided in terms of melting point and polarity.

If there is an oily constituent that is decided to be used, other oils should have different polarity and melting points from the oil. To produce stable emulsion against temperature and deterioration, at least two oils that mutually differ in polarity and melting point should be used. One of the oils should be solid at the room temperature, and the other should be liquid and better have a melting point below 0°C. For reference, hydrocarbons have the lowest polarity, followed by waxes, vegetable oils and fats, esters, higher alcohols, and fatty acids, in ascending order.

For example, let us assume producing a fluid cream that contains olive oil, which is liquid vegetable oil at the room temperature and has intermediate polarity. Paraffin wax can be used as the nonpolar oil, and solid myristyl alcohol and liquid octyldodecanol may be chosen as the highly polar oil constituents.

Adding Polymers

Polymers are frequently added to ensure the stability of emulsion and improve the feel of use. Thorough stability tests should be conducted because polymers may destabilize emulsions.

Widely used stable polymers for anionic, fatty-acid, and nonionic emulsions include carboxyvinyl polymers, polyacrylic acids, and xanthan gum.

For cationic creams, cationic guar gum, cationic cellulose, and hydroxyethyl cellulose are widely used as stabilizers. Various kinds of polymers are used in hair care cosmetics not only for stabilization but also for improving the feel of use and setting performance. There are also polymeric emulsifiers such as acrylates/C10–30 alkyl acrylate crosspolymer, polyacrylamide, hydroxyethyl acrylate/sodium acryloyldimethyl taurate copolymer, and dimethylacrylamide/ethyltrimonium chloride methacrylate copolymer.

3.2 Factors Affecting Viscosity and Stability and Preventive Measures

When an emulsion loses stability, it undergoes changes in properties such as separation, coagulation, sedimentation, viscosity change, and loss of uniformity. The major cause is inappropriate formulation, but temperature and light may also destabilize emulsions. Emulsions also separate by lowered viscosity at high temperatures and by aging at the room temperature.

3.2.1 Separation and Viscosity Increase at High Temperatures

Separation at high temperature occurs because the low-melting-point oily constituent becomes highly fluid liquid and allows light-weight emulsion particles to float. This can be somewhat prevented by adding ingredients of high melting points. Low HLB of nonionic surfactant, which results in insufficient temperature resistance, and effects of salt concentration are also possible causes, which can be solved by changing the percentage of the nonionic surfactants or alkyl group.

3.2.2 Separation and Viscosity Increase at the Room Temperature

3.2.2.1 Viscosity Increase

Viscosity increases with time are frequently observed in creams and cause changes in appearance and fluidity.

Viscosity increases (or thickening) are prevented by adding branched-chain higher alcohols (such as hexyldodecanol, isostearyl alcohol, and octyldodecanol) and nonpolar liquid hydrocarbons (such as liquid paraffin, liquid isoparaffin, and squalane). Branched-chain higher alcohols inhibit the crystallization of cetanol and keep the oily component fluid. Nonpolar liquid hydrocarbons, which are little affinitive to water, impede formation of hydrogen and ionic bonds. Increases in viscosity can be prevented by adding a small amount (5–20% of the oily component) of the compounds. At least one of surfactants should have a setaryl, lauryl, oleyl, or stearyl group. Use of surfactant that has an alkyl group, either alone or combined with another surfactant, keeps fluidity. Anionic and cationic surfactants should be combined with surfactants that have an OH group or POE.

For example, adding 0.5% of PEG-2 oleammonium chloride and sodium laureth sulfate improves the affinity to water and suppresses viscosity increases.

3.2.2.2 Controlling Viscosity Decreases (Or Thinning) with Time

(a) Reduce the amount of straight-chain higher alcohols, such as cetanol and stearyl alcohol, and add branched-chain higher alcohol, such as isostearyl alcohol and octyldodecanol
(b) Add liquid hydrocarbon such as squalane
(c) Replace part of surfactants to those that have lauryl, oleyl or isostearyl alkyl group
(d) Add water-affinitive ionic surfactant

3.2.2.3 Controlling Viscosity Decreases at High Temperatures

(a) Add polymers, which are little affected by viscosity changes
(b) Use surfactants that have cetyl or stearyl groups
(c) Avoid the use of surfactants that have isostearyl or oleyl groups

3.2.2.4 Separation

Separation with time at the room temperature occurs when the surfactants are not appropriately combined. The combination of surfactants needs to be revised.

Inappropriate formulation is another cause of separation. The polarity distribution of water, surfactants and oils must be revised to improve stability. Separation can be prevented by (1) combining water, glycol and ionic and nonionic surfactants so that all components including polar and nonpolar oils are uniformly distributed, and (2) adding a small amount of a substance that has an intermediate polarity of the major components.

3.2.3 Effects of Salts

Salts (acids and alkalis) added for pH adjustment may cause the emulsion to separate and/or decrease in viscosity.

3.2.3.1 Viscosity Decreases by Salts

Emulsions are viscous because hydrogen bonds are formed between emulsion particles and water molecules. Salts break the hydrogen bonds and cause drops in viscosity.

It is effective to increase the amount of nonionic surfactants in relation to ionic surfactants to prevent viscosity from lowering. Nonionic surfactants, which do not form ionic bonds, are less sensitive to salts compared to ionic surfactants. However, nonionic surfactants may adversely affect the high temperature stability because the clouding point is lowered by the presence of salts. The high temperature stability can be secured by adding an ionic surfactant that has a short alkyl group, which shows small salting-out effect. Diluting and lowering the concentration of the salt solution and adding the solution at low temperature are sometimes effective for preventing viscosity decreases that occur during salt addition.

3.2.3.2 Separation by Salts

Separation of an emulsion occurs when the amount of surfactants is insufficient or the HLB is inappropriate. Increasing the amount and raising or lowering HLB should be tried.

When a large amount of salts is to be added, it is effective to add an ionic surfactant that has a relatively short alkyl group. The key is to increase the percentage of nonionic surfactants.

3.2.3.3 Fluid Emulsions

(a) Replace part of surfactants to those that have lauryl or oleyl alkyl group
(b) Increase the amount of surfactants in relation to oils
(c) Add branched-chain higher alcohol
(d) Add liquid hydrocarbon to the oily component
(e) Increase the mutual solubility among oils. Add liquid esters

3.2.3.4 To Improve the Texture and Luster of Creams

(a) Use potassium hydroxide and triethanolamine as the counter ions of anionic surfactants and neutralizers
(b) Increase the amount of surfactants
(c) Increase nonionic surfactants of high HLB
(d) Increase the percentage of liquid oils in the oily component
(e) Increase the mutual solubility among oils. Add liquid esters
(f) Use potassium hydroxide and triethanolamine instead of sodium hydroxide for the alkali agent in soap emulsion

3.2.4 *Factors Causing Destabilization of Emulsions*

Creams may undergo changes in external appearance and/or crystal formation. Cationic and ionic components may react, gradually form complexes, and destabilize the cream. Polymer components may induce emulsion particles to coagulate at the functional groups. Polymers are added to stabilize emulsions and control fluidity, but they sometimes act the reverse.

Crystallization and reduced viscosity at low temperatures cause pearl appearance and occur when only solid oils are used, the amount of straight-chain alkyl group is large, the compatibility between oil components is low, etc. When only solid oils are used, the oils form crystals, which grow in size. Molecules that have straight-chain alkyl groups orient themselves and form crystals. A combination of liquid paraffin and stearyl alcohol, which are little compatible due to the large difference in polarity, is prone to crystallization. This can be solved by adding a small amount of liquid ester, which has an intermediate polarity of the two. Viscosity reduction occurs because crystals are formed and cause the oily components to separate from the emulsifier, freeing emulsion particles. This is sometimes recovered by raising the temperature to the room temperature.

3.2.4.1 Combinations Prone to Destabilization

(a) Anionic + cationic surfactants
(b) Polymer + polymer
(c) Ionic surfactant + polymer
(d) High molar nonionic surfactant + polymer

3.2.4.2 Ways to Enhance Emulsion Stability

(a) Raise or lower the HLB to the optimum HLB level
(b) Combine ionic surfactant with POE surfactant
(c) Add liquid ester

3.2.4.3 When Emulsification Fails

(a) Change the HLB of the surfactant
(b) Increase low-HLB or high-HLB surfactant
(c) Replace all or part of the surfactant to those that have lauryl or oleyl alkyl group or
(d) Improve the compatibility between polar and nonpolar oily components, and add liquid ester

3.3 Emulsification

Once the formulation of an emulsion is designed, the conditions of preparations should be investigated. The actual conditions used in industrial production may differ from those used in laboratory. For successful emulsification even in industrial scale, it is essential to investigate the method and order of adding the phases and the device and conditions of emulsification.

3.3.1 Emulsification Method

The mixing method of the oil and aqueous phases, i.e., whether to add the aqueous phase into the oil phase or to add the oil phase into the aqueous phase, is sometimes determined by the facility and may affect the emulsifiability and the quality of emulsion. Formulation should be designed in general so that emulsions of the same properties can be produced by either method (by adding the aqueous phase into the oil phase or the other way around).

The timing of adding an emulsifier for mixing oils and water is summarized in Table 3.3.

Table 3.3 Emulsification method and timing of adding emulsifier [1]

Emulsification method	Characteristics
Nascent soap method (in situ soap method)[a]	Higher fatty acids and alkali are mixed to form soap (emulsifier)
Emulsifier is added to the oil phase	A widely used method for cosmetics. It is advantageous in terms that emulsion particles are formed immediately after adding oil to water
Emulsifier is added to the aqueous phase	Requires strong agitation and is not appropriate for cosmetics. A large amount of surfactant sufficient for surrounding oil particles is needed for emulsification
Phase-inversion emulsification	Emulsifier is added to both the oil and aqueous phases. Stable emulsions are formed
	[a]By using the clouding point of nonionic surfactants, the method produces W/O emulsions at a high temperature, which turn into O/W emulsions when cooled. The temperature at which W/O turns into O/W is called the phase inversion temperature, and the emulsion of the higher phase inversion temperature is more stable. Theoretically, the emulsions do not separate at temperatures below the phase inversion temperature

[a][1]

Table 3.4 Emulsification devices and characteristics

Emulsification device	Structure and characteristics
Paddle mixer	Appropriate for emulsification of cosmetics. It emulsifies without shearing the components
Disperser	The strong shearing force is not appropriate for making emulsions. It is used for dispersing and dissolving polymers
Homomixer	Very effective for emulsifying cosmetics. It should be used only for the initial stage of emulsification. Long use destroys emulsions and causes destabilization and changes in viscosity
Propeller mixer	Appropriate for dissolving substances. It is used for dissolving substances and not for emulsification; and constant production of emulsions of the same appearance and viscosity is difficult even under constant emulsification conditions

3.3.2 Emulsification Device

Emulsification devices include paddle mixers, dispersers, homomixers, and propeller mixers. Each has own characteristics, and the device to use should be decided based on the purpose of manufacture. See Table 3.4.

3.3.3 Conditions of Emulsification

3.3.3.1 Emulsification Conditions and Viscosity

Emulsions are produced by using the chemical characteristics of surfactants and applying appropriate physical energy to the oil and aqueous phases. Factors affecting emulsification include the duration and temperature of emulsification and stirring speed. Stable emulsions can be produced by adjusting the factors correctly. The points to note during emulsification are summarized below.

Tips in Controlling Emulsification Conditions for Preventing Viscosity Changes

(a) Durations of emulsification and cooling process

Long stirring causes drops in viscosity

Change the cooling speed by season and control the temperature and flow rate of the coolant so that the emulsion is cooled for the same duration regardless of season

Avoid long stirring. (Long stirring may destroy the emulsion)

(b) Emulsification temperature
Heat both the aqueous and oil phases to the predetermined temperature. Low temperatures may cause large and coarse emulsion particles. (Heating reduces the viscosity of the oils and facilitates mixing.)

(c) Stirring speed
Do not increase the stirring speed thoughtlessly. It reduces the viscosity of the emulsion and destabilizes it. → Rapid stirring shears the mixture and reduces viscosity

(d) Methods and order of adding components
For o/w, the oil phase is added to the aqueous phase. For soap emulsification, the aqueous and oil phases are mixed first and then the other components are added.

3.3.3.2 Use of Constant Emulsification Conditions

Conditional factors should be minimized to simplify operation. Constant conditions should be used to always produce emulsions of the same appearance and viscosity and minimize quality fluctuations among lots. Tips for making emulsification conditions constant are summarized below.

Checklist for Assuring Constant Conditions of Emulsification

1. Mixing method of components
   Decide the order and speed of mixing the component
   → Wrong order may result in poor quality
2. Emulsification temperature
   Record the temperature of the oil and aqueous phases
   Adjust the temperature within the ±3 °C range of the predetermined temperature (usually 80 °C or higher)
   → Low temperature may result in failure of emulsification
3. Stirring speed at the initial stage of emulsification
   Control the stirring speed and duration (mainly for using homomixers)
   → The stirring speed and duration affect the texture and viscosity of the emulsion
4. Temperature control of coolant
   Record the temperature of the coolant at the inlet and outlet
   Devise methods so that the temperature is constant regardless of season (by controlling the flow rate)
   → Constant temperature must be ensured to assure constant cooling time

5. Stirring speed
   Assure constant stirring speed. Record the revolution
   → Varied stirring speed causes large variation in viscosity
6. Duration of emulsification
   Record the hour of starting and finishing each process [Example: Addition of coloring agent stated at 10:32. Addition of the coloring agent finished at 10:47. (Do NOT state like "Addition started 21 minutes after emulsification".) The records help judge whether the procedures were appropriately performed and controlled.]
   → Long stirring reduces viscosity and destabilizes the emulsion
7. Adding additives
   Add perfumes, coloring agents, salts, etc. always at the same temperature, dilution ratio and speed
   Always stir for the same duration
   → Thick solutions of coloring agents and salts destroy emulsions
   → Rapid addition also destroys emulsions
   If addition of salts causes sharp reductions in viscosity,
   → Slow down the adding speed
   → Reduce the concentration of the salt solution
   → Lower the temperature during addition
   Do not add additives during emulsification and at high temperatures
   Avoid adjusting pH after emulsification
8. Stirring at low temperatures
   Emulsions need not to be cooled to the room temperature
   Do not stir in excess, but stop stirring at the predetermined time
   → Excessive stirring causes drops in viscosity

3.3.3.3 Items to Investigate for Scaling up from the Laboratory to Production

Different points from those in the laboratory should be noted for full production. Scaling-up may require big changes in stirring conditions and temperature control. Here, points to note and solutions are outlined.

Stirring Speed

Set the stirring efficiency of the laboratory device at the level of the emulsification device used for production.

Increase the revolution of the laboratory device so that the stirring speed is the same as the production device.

Calculate the circumferential speed from the revolution of the paddle and the diameter of the container, and adjust the speed.

$$\text{Circumferential speed} = 2\pi r \times \text{number of revolutions}$$

* The actual stirring speed needs adjustment even when the laboratory and production devices have the same structure. The same stirring speed may not be achieved when it is stirred long.

Cooling Speed

Do not rapidly cool the emulsion in the laboratory. In the laboratory, particularly care should be taken on temperature control of the coolant, and flow rate control of the coolant for products prone to rapid cooling.

Reconfirming the Emulsification Device and Stirring Speed

Investigate whether the presently used method is appropriate for the production of the emulsion.

Design and formulate new products by fully understanding the production facilities.

Reference

1. Hiroshi H (1970) Oils, fats and surfactants for cosmetics (in Japanese). Saiwai, Shobo

Chapter 4
Sensory Properties of Cosmetics

Abstract Sensory evaluation is indispensable for objective assessment of products. Evaluation methods should be established for objectively understanding the impressions of users and also for deciding the words and phrases to be used for advertising the new products. Quantitative evaluation by using instruments is a field in which future progress is expected, but does not yet always reflect the actual impressions felt by users and cannot substitute sensory evaluation by people.

Knowing the feels given by each ingredient of cosmetics helps designing formulations. Especially the alkyl groups of oils and surfactants affect the feel of use of the product directly. The properties of the alkyl groups are the key factor in designing the feel of cosmetics, and we should understand of the chemical and physical properties and know the feel given by each ingredient.

Keywords Alkyl groups • Evaluation sheets • Sensory properties

4.1 Expressions of Sensory Properties

Good cosmetics give the feel consistent with the purpose of use when applied on the skin, hair, etc. The feel experienced by users varies by person, and so the expressions they use vary. However, objective and correct evaluation of sensory properties is an indispensable process in development of cosmetics products. In this chapter, expressions of sensory properties are classified and analyzed to deepen our understanding on cosmetics materials.

H. Iwata and K. Shimada, *Formulas, Ingredients and Production of Cosmetics: Technology of Skin- and Hair-Care Products in Japan*, DOI 10.1007/978-4-431-54061-8_4,

Table 4.1 Expressions of sensory properties

	Major expression	Similar expressions
Positive impressions	Moisturizing	Oily, wet, heavy, moist
	Silky	Light, refreshing
	Smooth	Slippy, slippery, velvet, flowing
	Soft	Pliable, flexible
	Persistent	Adhering, coated
Negative impression	Dry	Hard, tight, stiff, coarse, dull

4.1.1 Classification of Expressions

The senses given by cosmetics may vary by user. Users may also use various expressions to describe the feel. General positive impressions of using cosmetics include "moisturizing," "smooth," "silky," and "soft."

"Dry" is a negative expression. "Persistent" is used for both positive and negative impressions, but basically as a positive impression in this book.

There are various similar expressions besides these six. Adverbs such as "very," "strongly," "rather," and "slightly" are widely used combined with the expressions (Table 4.1).

4.1.2 Favored Senses

4.1.2.1 By Purpose of Use

Senses favored by users vary depending on the category of cosmetics, body part on which it is used, and season.

For example, facial and hand creams have similar compositions, but the sense favored by users differs. Preferred hand creams are those that make the skin "smooth" but not "slippery." On the other hand, users demand for "moisturizing" facial creams in general.

The impressions felt from using skincare cosmetics are affected by seasonal temperature and humidity changes, and so the demanded sensory characteristics change by season (Table 4.2).

4.1.2.2 By User

Favored senses may also vary by user. For example, there is a big difference in favored sense particularly of rinse-off hair care products between home users and professional beauticians. Home users prefer shampoos and conditioners that are quick to rinse off the "smooth (lubricious)" feeling. On the other hand, professional beauticians evaluate products that long retain the "smooth (lubricious)" feeling.

Table 4.2 Categories of cosmetics and sensory properties favored by users

Body part	Method of use	Category	Favored senses
Hair	Rinse off	Shampoos, conditioners	Smooth, moisturizing
	Leave on	Leave-on hair care products	Slippy, silky
Face	Rinse off	Cleansing oils	Moisturizing
	Leave on	facial toner, creams	Summer: silky, winter: moisturizing
Body	Rinse off	Body soaps	Smooth, moisturizing
	Leave on	Body milks	Light, silky

4.2 Sensory Evaluation

It is indispensable, although it involves very difficult tasks, to establish sensory evaluation methods for objectively understanding the impressions of users, and also for deciding the words and phrases to be used for advertising the new products.

Measuring instruments are also used today to collect objective numerical data on friction, bending strength, etc. Quantitative evaluation by using instruments is a field in which future progress is expected. However, the results calculated from collected numerical data are often different from the actual impressions felt by users. At the moment, human sensory evaluation is superior to measuring instruments in perceiving the entire and versatile sensory characteristics of cosmetics products.

4.2.1 Points to Note in Human Sensory Evaluation

Personal impressions are somewhat included in all evaluations conducted by people. Understanding the trends of personal impressions helps correct evaluation.

Knowledge on the formulation may affect the evaluation. Evaluation and review by members of the institute who are not engaged in the development of the product are more objective than those by people engaged in the development, and are thus informative and useful. Building such a relation is recommended because it also helps you to obtain advice afterwards.

The most important and final sensory evaluation is the one conducted by users, including professional beauticians and home users. Answers can be obtained from a large number of people and can be statistically processed. Use of simple formats to guide the respondents is recommended to prevent diversities in answers and expressions.

Each product should be evaluated separately for each process of use: for example, at application, at rinsing, when the hair is wet, during towel-drying, during blow-drying, and after finishing. Examples of evaluation formats for cleansers, shampoos, conditioners, and leave-on products are given below.

Sensory evaluation sheet (cleansers)

When applied on the skin			
Lathering and bubble texture	Good	Normal	Bad
Lubrication (smoothness)	Good	Normal	Bad
Softness	Good	Normal	Bad
Irritation[a]	Not irritating		Irritating
While rinsing			
Long-lasting sliminess	Long lasting	Normal	Not lasting
Feel when rinsing	Good	Normal	Bad
Lubrication (smoothness)	Good	Normal	Bad
Irritation[a]	Not irritating		Irritating
When toweling			
Moisturized/silky	Moisturized	Intermediate	Silky
Lubrication (smoothness)	Good	Normal	Bad
After cleansing			
Moisturized/silky	Moisturized	Intermediate	Silky
Lubrication (smoothness)	Good	Normal	Bad
Feel of dryness	None	Normal	Feel dry
Stiff face	None	Normal	Stiff face

[a] Instead of "irritation," more concrete expressions may be used, such as "prickling," "painful," and "hurting"

Sensory evaluation sheet (shampoos)

When applied on the hair			
Lathering and bubble texture	Good	Normal	Bad
Lubrication (smoothness)	Good	Normal	Bad
Softness	Good	Normal	Bad
Irritation[a]	Not irritating		Irritating
While rinsing			
Long-lasting sliminess	Long lasting	Normal	Not lasting
Finger pass when rinsing	Smooth	Normal	Not smooth
Lubrication (smoothness)	Good	Normal	Bad
Coarseness, roughness	None	Normal	Coarse, rough
Irritation[a]	Not irritating		Irritating
When towel-drying			
Moisturized/silky	Moisturized	Intermediate	Silky
Lubrication	Good	Normal	Bad
Smoothness	Good	Normal	Bad
Coarseness, roughness	None	Normal	Coarse, rough

[a] Instead of "irritation," more concrete expressions may be used, such as "prickling," "painful," and "hurting"

Sensory evaluation sheet (conditioners)

When applied on the hair			
Spread to the hair	Good	Normal	Bad
Lubrication	Good	Normal	Bad
Sliminess	Good	Normal	Bad
Softening the hair	Softened	Normal	Not softened

(continued)

Sensory evaluation sheet (conditioners)

While rinsing			
Lubrication	Good	Normal	Bad
Long lasting sliminess	Long lasting	Normal	Not lasting
Coarseness, roughness	None	Normal	Coarse, rough
When towel-drying			
Moisturized/silky	Moisturized	Intermediate	Silky
Lubrication	Good	Normal	Bad
Soft/hard	Soft	Intermediate	Hard
When blow-drying			
Hair unites/spreads	Unites	Intermediate	Spreads
Lubrication	Good	Normal	Bad
Soft/hard	Soft	Intermediate	Hard
Moisturized/silky	Moisturized	Intermediate	Silky

[a] Instead of "irritation," more concrete expressions may be used, such as "prickling," "painful," and "hurting"

Sensory evaluation sheet (leave-on hair care products)

When applied on the hair			
Spread to the hair	Good	Normal	Bad
Lubrication	Good	Normal	Bad
Sliminess	Good	Normal	Bad
Vanishing (whitening)	None	Normal	Vanishing
Tackiness	None	Normal	Tacky
When blow-drying			
Hair unites/spreads	Unites	Intermediate	Spreads
Lubrication	Good	Normal	Bad
Soft/hard	Soft	Intermediate	Hard
Moisturized/silky	Moisturized	Intermediate	Silky
Uniformity (from the root to the tip)	Good	Normal	Bad
Tackiness	None	Normal	Tacky
Luster	Lustrous	Normal	Not lustrous

Sensory evaluation sheet (skincare products)

When applied on the skin			
Spread on the skin	Good	Normal	Bad
Lubrication	Good	Normal	Bad
Smoothness	Good	Normal	Bad
Vanishing (whitening)	None	Normal	Vanishing
Tackiness	None	Normal	Tacky
Irritation[a]	Not irritating		Irritating
After application			
Smoothness	Good	Normal	Bad
Moisturized/silky	Moisturized	Intermediate	Silky
Coarseness	None	Normal	Coarse
Tackiness	None	Normal	Tacky

[a] Instead of "irritation," more concrete expressions may be used, such as "prickling," "painful," and "hurting"

4.3 Sensory Properties of Alkyl Groups

It is very important to know the feels given by each ingredient of cosmetics. Especially the alkyl groups of oils and surfactants affect the feel of use directly.

The alkyl groups determine the melting point (the larger the molecular weight, the higher the melting point) and polarity (by the number of oxygen atoms and chemical structures, such as double bonds, side chains, and functional groups). The functional groups determine the compatibility and interaction with other ingredients and affect the viscosity, consistency, appearance, hardness, spread, and the feel of use. The sense and texture felt by users highly depends on the structures of alkyl groups. The properties of the alkyl groups are the key factor in designing the feel of cosmetics, and we should understand of the chemical and physical properties and know the feel given by each ingredient.

4.3.1 Sensory Properties of Oils

Oils have an alkyl group as the basic structure and include carbohydrates, which do not have an oxygen atom, esters and waxes, which have ester bonds, and higher alcohols and fatty acids, which have oxygen atoms at an end of the molecule. The alkyl and functional groups determine the sensory properties of oils.

4.3.1.1 Length of Alkyl Group and Sensory Properties

Short alkyl groups give a "coarse" touch, those of intermediate length give "lubricious" feeling, and long alkyl groups are "soft" and "moisturizing." Double bonds in alkyl chains increases "moisturizing" and "oily" feeling; and branched chains give "silky" and "light" touch.

Complicated molecular structures increase the "moisturizing" and "oily" feelings. Oils of larger molecular weights have higher melting points, are solids at higher temperatures, and thus give less "lubricious," less "silky," and less "smooth" touches (Table 4.3).

Waxes, which have higher melting points than oils, are "adhesive" and give a coated feeling. Ester bonds, hydroxyl groups, and carboxyl groups do not directly affect the feel of use, but affect the viscosity and consistency of products. Higher viscosities give more "smooth" and "soft" feeling, and low viscosities result in "lightness." Increase the content of liquid oils to produce "lubricious" skincare cosmetics, and increase the solid oil content for reducing the "lubricious" feeling.

Table 4.3 Alkyl group in glycols and feel of use

Alkyl group	Fatty acid higher alcohol	Higher alcohol	Feel of use
n-12	Laurate	Lauryl alcohol	Coarse
n-14	Myristate	Myristyl alcohol	Silky, lubricious
n-16	Palmitate	Cetyl alcohol	Moisturizing
n-18	Stearate	Stearyl alcohol	Moisturizing
n-20	Arachidic acid	Arachidyl alcohol	
n-22	Behenic acid	Behenyl alcohol	Soft
e-18	Oleic acid	Oleyl alcohol	Oily, moisturizing
e-18"	Linoleic acid		
iso-16		Hexyldecanol	Light, lubricious
iso-18	Isostearic acid	Isostearyl alcohol	Lubricious
iso-20		Octyldodecanol	Lubricious
oh-18	Hydroxystearic acid		

4.3.2 Sensory Properties of Surfactants

The hydrophilic groups of anionic surfactants are either carboxyl, phosphate, sulfonate, or sulfate. The number of oxygen and sulfur atoms in the groups determines the hydrophilicity of the surfactant.

Cationic surfactants have trimethyl ammonium as the hydrophilic group, and the feel of use is determined by the length of the alkyl group. As described in the section of cationic surfactants, lauryl, which has a short alkyl group, gives a coarse feeling. The longer the alkyl group (cetyl < stearyl < behenyl), the more moisturizing feel it gives.

Amphoteric surfactants are widely used in shampoos, facial cleansers, and body soaps. Most of the surfactants have a lauryl or cocoyl group, and the feel of use is determined by the structure of the hydrophilic group. Those that have a simple structure and consist of a straight chain such as lauryl betaine give a light texture; and those that have a complicated structure such as sodium cocoamphoacetate give a non-slippery moisturizing touch.

Nonionic surfactants may give coarseness depending on the length of polyoxyethelene. As in higher alcohols, the sensory properties are affected by the length of the alkyl group, double bonds and side chains but much slightly compared to cationic surfactants.

Highly hydrophilic surfactants are easily rinsed off from the skin and hair; and those of low hydrophilicity tend to remain long. Polyoxyethylene sodium lauryl ether sulfate, which has a sulfate group, makes silky shampoos in general; and amino acid surfactants are known to make moisturizing shampoos.

Table 4.4 Sensory properties of glycols

	Feel of use	
Constituent[a]	Skin	Hair
Diglycerin	Smooth, moisturizing	Rather sticky
Glycerin	Rather moisturizing	Sticky
1,3-Butyleneglycol	Moisturizing	Rather sticky
Isoprene glycol	Rather silky	Light and refreshing
Dipropylene glycol	Soft, smooth, very silky	Refreshing
Proprene glycol	Very silky	Very silky, very refreshing

[a] 8% solution

4.3.3 Sensory Properties of Glycols

The sensory properties of glycols, which are used as moisturizers, are also determined by the molecular structure, particularly by the number and positions of the hydroxyl groups and the molecular weight.

For example, let us compare glycerin, 1,3-butylene glycol and proprene glycol, which are most widely used glycols in cosmetics. As in facial toners, glycerin, which has three carbons and three hydroxyl groups, gives a "heavy (very moisturizing)" feeling. 1,3-Butylene glycol, which has three carbons and two hydroxyl groups, gives a "moisturizing" texture. Proprene glycol (three carbons and two hydroxyl groups) is "light."

Isoprene glycol, which has five carbons and one more methyl groups than 1,3-butylene glycol and is thus more hydrophobic, gives a "lubricious" touch. Dimer glycols such as dipropylene glycol and diglycerin increase viscosity. The former adds softness and the latter increases the oily and moisturizing feeling. It is important to know the basic sensory properties of glycols because they affect the feel of use of cosmetics products.

The textures given by glycols to facial toner and hair mist are shown in Table 4.4.

Part II
Formulas and Productions of Cosmetics

Chapter 5
Practice of Designing Cosmetics Formulations

Abstract This chapter describes the practice of designing cosmetics formulations and gives examples of actual formulations. Toilet soaps have long been manufactured by saponifying natural oils and fats with alkalis, such as sodium hydroxide. Today, they are mainly produced by neutralizing fatty acids. Shampoos mainly consist of anionic, amphoteric and cationic surfactants, thickeners, and bubble stabilizers. The properties and combinations of the ingredients determine the viscosity and the feel of use of the product and thus should be carefully investigated. Hair conditioners, rinses, treatments, and packs consist of similar constituents, but hair treatments and hair packs also contain proteins, amino acids, moisturizing ingredients, and/or other efficacious constituents in general to repair damage. Skin creams should be designed so as to achieve the aimed sensory characteristics, i.e., for making moisturizing or light cream. The sensory characteristics of creams are determined by the oily components, emulsifiers, polymers, glycols, product form, and viscosity. The sensory characteristics and viscosity can be estimated from the alkyl groups of the oily components and surfactants. Leave-on hair care cosmetics are applied directly on hair and are not rinsed off, and thus the sensory characteristics of the constituents are directly felt. It is thus important to have thorough knowledge of the sensory characteristics of the ingredients to be combined such as oils, glycerols, and polymers. Cosmetics mainly consisting of oils are produced by combining carbohydrates, vegetable and animal oils and fats, waxes, esters, higher alcohols and nonionic surfactants, and adding coloring agents, scenting agents, and efficacious ingredients. Facial toner should be formulated so as to moisturize, soften, firm up, brighten, and smoothen the skin and give a moisturized feeling.

Keywords Facial toner • Hair conditioners • Leave-on • Rinse-off • Shampoos • Skin creams • Soaps

H. Iwata and K. Shimada, *Formulas, Ingredients and Production of Cosmetics: Technology of Skin- and Hair-Care Products in Japan*, DOI 10.1007/978-4-431-54061-8_5,

5.1 Soaps and Cleansers

Soaps and cleansers mainly consist of soap and include toilet soaps, body soaps, and facial cleansers. Toilet soaps have long been manufactured by saponifying natural oils and fats with alkalis, such as sodium hydroxide. Today, they are mainly produced by neutralizing fatty acids.

5.1.1 Combinations of Ingredients and Products

Fatty acids and alkalis should be selected based on the properties of the product. For toilet soaps and powder soaps, sodium salts of lauric acid, palmitic acid, stearic acid, and oleic acid are used. Potassium salts of lauric acid, myristic acid, and palmitic acid are used for creamy and paste products, such as facial cleansing creams.

The properties of soaps are determined by the fatty acid composition and alkali agent used. Sodium salts of long chains of saturated fatty acids produce solid soaps, and short chains of potassium salts of unsaturated fatty acids become creamy or liquid. Soaps are governed by the melting point of the fatty acid component and the number of carbons in the alkyl group of the fatty acid. Large number of carbons results in solids, and small number of carbons produces creams and liquids. However, oleic acid has a double bond in the middle of the molecule, and its sodium and potassium salts produce pastes and transparent viscous liquid, respectively.

5.1.2 Toilet Soaps

Toilet soaps are produced by either saponification or neutralization of fatty acids. The former involves saponifying 80% tallow and 20% coconut oil with sodium hydroxide, extracting neat soap by salting out with a saline solution, and drying. The latter involves neutralizing the fatty acids with sodium hydroxide to produce neat soap.

There are two methods for drying the neat soaps to produce solid soaps: drying in molds and mechanical kneading. Mechanically kneaded soaps have low moisture content and are suitable for bath, and those dried in molds are suitable for facial cleansing.

Mold drying involves placing the neat soap in molds and drying. The Japanese Industrial Standards (JIS) prescribes the moisture content of the soaps to be 28% or less.

Mechanical kneading involves vacuum drying the neat soaps into pellets, kneading thoroughly by using rollers, etc., heating and squeezing into bars, and forming the product shape by applying a die. The required moisture content under JIS is 16% or less (Prescription 5.1).

Prescription 5.1 Toilet soap (mold drying)

	Part	Ingredient	% (100 g)
1	A	Coconut acid (nv255)	20.00
2	A	Tallow fatty acid or palm oil fatty acid (nv198)	80.00
3	A	Glycerin	6.00
4	C	EDTA-2Na	0.20
5	C	Sodium citrate	0.50
6	B	Sodium hydroxide	17.57
	B	Purified water	44.00

Directions
1) Weigh 1–3, and heat to 80°C. (A)
2) Dissolve 6 in purified water. (B)
3) Mix A, B, 4 and 5. (A+B+C)
4) Spread in a mold and dry

Specifications
pH: 10.0–10.5 (1% solution)

5.1.3 Cream Soaps

Cream soaps such as facial cleansers mainly consist of potassium salts of lauric acid, myristic acid, palmitic acid, and stearic acid. The properties, such as the melting point, of the potassium salt of fatty acid depend on the fatty acid component. Therefore, it is difficult to maintain a certain level of viscosity over a wide temperature range. To keep a certain viscosity, two or more potassium salts of different fatty acids should be combined. The ratios and amounts of the fatty acids should also be carefully investigated.

The stability at each temperature range can be checked by differential thermal analysis (DTA) and differential scanning calorimetry (DSC). Higher alcohols, such as cetanol, glycerin, and PEG, are indispensable for improving stability, and amphoteric and anionic surfactants are also needed (Prescription 5.2).

5.1.4 Liquid Soaps

Almost nonviscous liquid soaps are produced by neutralizing coconut oil with potassium hydroxide in general (Prescription 5.3). Coconut oil contains at least 45–50% lauric acid and also caprylic acid, capric acid, myristic acid, and a small amount of oleic acid, and is the most suitable oil for producing liquid soaps.

Appropriate concentration is 10–30%. Concentrations above 30% result in gels. To prevent gelation of high concentration liquid soaps, 5–6% of glycerin or propylene glycol needs to be added. However, the gelation problem does not occur at high temperatures because the solubility of the potassium salts of long-chain fatty acids is low at low temperatures.

Prescription 5.2 Cleansing cream (for facial cleansing)

	Part	Ingredient	% (100 g)	% (100 g)
1	A	Lauric acid (nv280)	4.00	2.00
2	A	Myristic acid (nv220)	8.00	8.00
3	A	Stearic acid (nv210)	12.00	14.00
4	B	Potassium hydroxide	11.06	6.44
5	B	Sodium citrate	0.50	0.50
6	B	EDTA-2Na	0.20	0.20
7	B	Sodium gluconate	0.50	0.50
8	B	Glycerin	3.00	3.00
9	B	Polyethylene glycol	6.00	6.00
10	C	1,2-Pentanediol	1.00	0.20
11	C	Ethylene glycol distearate	3.00	2.00
12	C	Squalane	0.20	0.20
13	C	Cetanol		1.50
14	C	PEG-120 methyl glucose dioleate	2.00	
15	C	Cocamide DEA		2.00
16	C	Sodium methyl cocoyl taurate (30%)		6.00
		Purified water	to 100 g	to 100 g

Directions
1) Weigh 1–3, and heat to 70°C. (A)
2) Dissolve 4 to 9 in purified water. (B)
3) Mix A and B, and heat to 85°C. (A+B)
4) Add 10–16 and stir until uniform . (A+B+C)
5) Cool to 40°C, measure the pH, adjust the pH with KOH (10%) and add water

Specifications
pH: 9.5–10.2

Prescription 5.3 Transparent liquid soap (for hands, body and face. Also for foaming bottles)

	Part	Ingredient	% (100 g)
1	A	Coconut acid (nv255)	12.5
2	A	Glycerin	3.00
3	A	Sodium citrate	0.50
4	C	Betaine	0.10
5	C	PCA-Na	0.50
6	C	Dipotassium glycyrrhizinate	0.10
7	B	Potassium hydroxide (85%)	4.34
	B	Purified water	to 100 g

Directions
1) Dissolve 7 in purified water. (B)
2) Add 1–3 to B and heat to 80°C. (A+B)
3) Cool to 50°C and add 4–6. (A+B+C)
4) Adjust the pH with 10% KOH and add water

Specifications
pH: 9.8–10.5

Low pH destabilizes liquid soaps. The pH should be adjusted at 9.8–10.5. Low pH causes cloudiness at low temperatures and coloring and oxidative rancidity by aging.

The viscosity of liquid soaps cannot be adjusted just by adjusting the concentration of the soap but needs other ingredients to be added. Soaps that mainly consist of potassium salts of fatty acids produced from coconut oil or palm kernel oil are slightly viscous in general. To increase the viscosity, it is necessary to add methyl cellulose, PEG-120 glucoside dioleate, etc. Coconut diethanolamide and coconut monoethanolamide, which are widely used for thickening shampoos, are ineffective in thickening liquid soaps. Long-chain fatty acids, such as palmitic acid and stearic acid, cannot be used for adjusting viscosity because they cause clouding and separation at room temperature.

5.1.5 Viscous Soaps

Transparent soaps are produced by adding ethanol, glycerin, sugar, etc. to soap materials and drying. They contain more sodium laurate than ordinary toilet soaps and thus lather better and are more soluble to water. Therefore, they are suitable for facial uses than for bath (Prescription 5.4).

Prescription 5.4 Transparent viscous liquid soap

	Part	Ingredient	% (100 g)	% (100 g)	% (100 g)
1	A	Coconut acid (nv255)	18.00	15.00	15.00
2	A	Oleic acid	2.00		
3	A	Citric acid	1.00	1.00	1.00
4	A	Glycerin	3.00	3.00	3.00
5	B	Potassium hydroxide (85%)	7.05	5.68	5.68
6	B	Purified water	24.00	24.00	24.00
7	C	EDTA-4Na	0.20	0.20	0.20
8	C	Sodium gluconate	1.00	1.00	1.00
9	D	PEG-120 methyl glucose dioleate			2.00
10	E	Hydroxy methyl propyl cellulose		0.40	
	E	Purified water	to 100 g	to 100 g	to 100 g

Directions
1) Dissolve 5 in purified water. (B)
2) Add 1–4 to B and heat to 80°C. (A+B)
3) Dissolve 10 in purified water. (E)
4) Dissolve 7–9 in A+B. (A+B+C+D)
5) Cool to 45°C and add E and water. (A+B+C+D+E)

Specifications
pH: 9.8–10.5

5.1.5.1 pH and Free Alkali

Generally, soaps contain free alkalis exceeding the amount necessary for neutralizing the fatty acid component, which are used to adjust the pH at 9.5–10.5. The amount of free alkalis in soaps is prescribed by JIS. The measurement involves weighing 5 g of specimen, boiling to remove carbon dioxide, adding an ethanol solution of phenolphthalein indicator neutralized with potassium hydroxide, melting the specimen on a water bath, and titrating with the standard 0.1 N hydrochloric acid solution. The consumption of the standard hydrochloric acid solution must not exceed 1 ml/g.

The JIS control value for free alkali is 0.1% or less.

5.1.5.2 Petroleum Ether Soluble Matter

Natural oils and fats contain unsaponifiable matters besides triglycerides. Natural oils and fats also contain unreacting oils because of the triglycerides. The amount of the unreacting oils, which is measured by extracting in petroleum ether, is called petroleum ether soluble matter in JIS and must not exceed 3% in toilet soaps. Scenting agents, squalane, cetanol, and other oily ingredients added to soaps are all petroleum ether soluble matters. These components do not affect the lathering and detergency of the soaps.

5.1.5.3 Effects of Chelating Agents

Fatty acid salts react with potassium and magnesium ions in water and form metal salts of fatty acids (soap scum), which are insoluble in water and reduce the lathering capacity of the soap. To prevent the reaction, chelating agents such as EDTA should be added to soaps. Citric acid also has a chelating capacity over a wide pH range and manifests synergistic effect with EDTA.

5.1.6 Cleansing Cosmetics

Cleansing cosmetics are for removing the filth from the face, hands, body, etc. They mainly consist of anionic surfactants and may also contain amphoteric surfactants and potassium salts of fatty acids (Prescriptions 5.5, 5.6, and 5.7).

The product forms vary from liquids to creams. Thickeners, such as ethylene glycol distearate, cetanol, other saturated higher alcohols, fatty acid alkanolamides, are added to produce viscous products. PEG-120 methyl glucose dioleate, hydroxy methyl propyl cellulose, ethylene glycol stearate, and ethyele glycol distearate may also be used as thickeners.

Prescription 5.5 Transparent viscous body soap (for pump bottles)

	Part	Ingredient	% (100 g)
1	A	Acrylic acid/(C10-30) alkyl acrylate cross polymer	0.40
2	A	Purified water	50.00
3	A	Glycerin	3.00
4	A	EDTA-2Na	0.10
5	B	BG	2.00
6	B	1,2-Pentanediol	0.80
7	B	Methyl paraben	0.20
8	B	Propyl paraben	0.05
9	C	Sodium methyl cocoyl taurate (30%)	18.00
10	C	Cocamide propyl betaine (30%)	12.00
11	C	Sodium cocoamphoacetate (30%)	4.00
12	C	Cocamide DEA	2.00
13	C	10% Potassium hydroxide	1.60
		Purified water	to 100 g

Directions
1) Disperse 2 into 1 uniformly. Add 3 and 4 and heat to 60°C. (A)
2) Heat 5–8, and homogenize. (B)
3) Add B to A, add 9–12, stir until uniform and neutralize with 13. (A+B+C)
4) Cool to 40°C, and add water

Specifications
pH: 6.0–7.5

Ingredient
*1 Pemuren TR-2 (Lubrisol Advanced Materials)

Prescription 5.6 Anionic-amphoteric facial cleansing foam (for foam bottles)

	Part	Ingredient	% (100 g)
1	A	EDTA-2Na	0.10
2	A	Sodium benzoate	0.20
3	B	BG	3.00
4	B	1,2-Pentanediol	0.90
5	B	PEG-20 sorbitan cocoate	0.40
6	B	Scenting agent	0.20
7	C	Sodium methyl cocoyl taurate (30%)	18.00
8	C	Cocamide propyl betaine (30%)	12.00
9	C	Sodium cocoamphoacetate (30%)	4.00
10	C	Cocamide DEA	2.00
11	C	Sodium dilauroamidoglutamide lysine (30%)	0.10
12	C	Glycerin	3.00
	A	Purified water	to 100 g

Directions
1) Dissolve 1 and 2 in purified water. (A)
2) Mix 3–6, and homogenize. (B)
3) Add 7–12 into A, add B, and add water. (A+B+C)

Specifications
pH: 6.0–7.5

Prescription 5.7 Liquid facial soap (for pump bottles)

	Part	Ingredient	% (100 g)
1	A	Potassium hydroxide	1.10
2	A	Citric acid	1.00
3	A	EDTA-2Na	0.30
4	A	Sodium benzoate	0.20
5	B	Potassium cocoate solution (40%)	20.00
6	B	Sodium methyl cocoyl taurate (30%)	8.00
7	B	Cocamide propyl betaine (30%)	6.00
8	B	Cocamide DEA	3.00
9	B	Glycerin	3.00
10	C	BG	2.00
11	C	PEG-20 sorbitan cocoate	0.40
12	C	Scenting agent	0.20
13	C	Phenoxyethanol	0.50
14	C	1,2-Pentanediol	0.80
		Purified water	to 100 g

Directions
1) Dissolve 1 into purified water, and dissolve 2–4. (A)
2) Mix 10–14, and homogenize. (C)
3) Add 5–9 into A, heat to 70°C, and stir until uniform. (A + B)
4) Cool to 50°C, add C and water. (A + B + C)

Specifications
pH: 9.2–10.2

5.2 Shampoos

Shampoos mainly consist of anionic, amphoteric and cationic surfactants, thickeners, and bubble stabilizers. The ingredients should be combined so as to fit the purposes in order to make good shampoo products.

It is essential to investigate the properties and combinations of the ingredients since they determine the viscosity and the feel of use of the product.

To ensure stability, use of stabilizers, preservatives and acidity regulators should also be examined. Today, many consumers prefer organic products, and there are demands for products that do not contain preservatives, silicone, synthetic surfactants or ingredients in the former list of prohibited materials.

5.2.1 *Purposes of Shampoo Formulation and Constituents*

The formulation of shampoo should be determined by examining the appearance, viscosity, lathering, detergency, safety, and feel of use. Most home-use shampoo products have pearl-like luster while those used by professional beauticians are transparent. The optimum viscosity is being easy to pour on a palm and still not falling out.

Smooth lathering is highly demanded, and creamy foams are usually favored. Products that are too viscous are not favored because they do not lather smoothly and produce sticky foams.

Detergency is not a big point of consideration because today the hair of most users is not greasy.

Safety is not a matter of big concern either except for eye irritation as long as safe ingredients are used.

5.2.2 Anionic Surfactants

Anionic surfactants are the major component of shampoos and give detergency and lathering performance. The anionic surfactants determine the safety (eye irritation), lathering performance, feel of use, and the viscosity of the product. They also influence the thickening effects of anionic surfactants, polymers, and nonionic surfactants, such as alkanolamide.

Most anionic surfactants have an alkyl group of lauryl, cocoyl, or palmitoyl. The surfactants are classified by the structure of the hydrophilic group into sulfonic-acid, sulfuric-acid, and carboxyl groups. The sulfonic-acid and sulfuric-acid groups are strongly anionic because they have three to four oxygen atoms and also sulfur atoms. On the other hand, the carboxyl group is less anionic.

Anionic surfactants form complexes with cationic polymers. Because the alkyl group is the same in all anionic surfactants, the hydrophilic group determines the lathering, feel of use, and the strength of the complexes. The amino acids and polyoxyethylene (POE) between the alkyl and hydrophilic groups of anionic surfactants influences the low-temperature stability and feel of use of the resultant shampoo products. POE prevents clouding at low temperatures, and those that have a nitrogen atom(s) such as amino acids improve the feel of use. Anionic surfactants that have sulfonic-acid or sulfuric-acid groups produce shampoos that lather well and give light and dry feeling, while those of the carboxyl group tend to give a moist feeling. The groups also determine the smoothness during rinsing.

Anionic surfactants are listed in Table 5.1. Select one or more from the list. The feel of use can be adjusted by combining two or more anionic surfactants.

Care should be taken in using anionic surfactants in shampoo formulations because the salt concentration in raw materials influences the low-temperature stability and viscosity of the products. At high salt concentrations, anionic surfactant shampoos are prone to clouding and separation at low temperatures; and the viscosity of the product tends to rise. In particular, when sodium lauryl sulfate is combined, addition of salt may raise viscosity. Anionic surfactants that have carboxyl groups tend to undergo clouding at low temperatures when the salt concentration is high.

Anionic surfactants are effective detergent at alkaline conditions, but shampoo products may be demanded to have low pH in their concept. Care must be

Table 5.1 Characteristics of anionic surfactants in shampoo formulations

		Viscosity	Lathering	Sensory
Glutamate	TEA-Cocoyl glutamate	Rather low	Rather low	Moist and heavy
N-acyl amino acid	Sodium cocoyl sarcosinate	Low	Normal	Light
Alanine	Sodium cocoyl methyl alanine	Low	Good	Soft
Alkyl ether carbonate	Sodium laureth sulfate	Intermediate	Rather low	Light
N-acyl taurine	Sodium cocoyl methyltaurate	Low	Good	Soft
Sulfosuccinate	Disodium laureth sulfosuccinate	Difficult	Rather low	Light
Sulfonate	Sodium C14-16 olefin sulfonate	High	High	Light
POE alkyl ether sulfonate	POE TEA-lauryl sulfonate	High	High	Light
Alkyl sulfate	TEA-alkyl sulfate	High	High	Light
PEG cocamide sulfate	PEG-3 cocamide MEA	Intermediate	Normal	Heavy

taken in selecting surfactants for products of low pH because there are anionic surfactants that easily undergo clouding and separation depending on their hydrophilic groups. Surfactants that have carboxyl groups are prone to clouding at low temperatures. On the other hand, those that have sulfonic-acid or sulfuric-acid groups are stable even in acids and perform effectively. Sodium lauryl sulfate is stable at low pH; and surfactants that have carboxyl groups should be used at pH of 6 or higher.

5.2.3 Amphoteric Surfactants

Amphoteric surfactants are the second largest component in shampoo formulations. Amphoteric surfactants are combined with anionic surfactants to improve the washing performance, lathering, anti-stimulus and sensory characteristics of shampoos. Amphoteric surfactants that have imidazolium give a persistent and non-slippery feeling. Widely used cocamidopropyl betaine gives a favorable silky touch. Lauryl betaine gives a light feeling and lathers well. Surfactants that have a carboxyl group make good combinations with cocamidopropyl betaine. When combined with sodium lauryl sulfate, sodium cocoamphoacetate results in the highest viscosity, followed by cocamidopropyl betaine and lauryl betaine in this order (Tables 5.2–5.4).

5.2.4 Cationic Polymers

Cationic polymers have big influences over the feel of use of shampoos. The most widely used cationic polymer is polyquaternium-10 (*O*-(2-hydroxy-3-(trimethy-lamminio)

Table 5.2 Difference in sensory characteristics of shampoo by anionic surfactants

Ingredients	% (100 g)
Prescription for sensory evaluation (1)	
Anionic surfactant	10.80
Cocamide propyl betaine	8.00
Lauramidopropyl dimethyl amine oxide	3.60
Cocamide MIPA	1.20
Methyl paraben	0.20
Polyquaternium-10	0.60
50% citric acid	0.20
Sodium benzoate	0.20
Purified water	to 100 g
Prescription for sensory evaluation (2)	
Anionic surfactant	10.80
Sodium lauroamphoacetate	8.00
Cocamide DEA	4.00
Methyl paraben	0.20
Polyquaternium-10	0.70
50% citric acid	0.20
Sodium benzoate	0.20
Purified water	to 100 g

Anionic surfactants evaluated	
1	Sodium cocoyl methyltaurate
2	PEG-3 Cocamide MEA
3	POE lauryl sulfonate
4	Ammonium lauryl sulfate
5	TEA-alkyl sulfonate
6	Disodium laureth sulfosuccinate
7	Sodium C14-16 olefin sulfonate
8	Sodium cocoyl sarcosinate
9	TEA-cocoyl glutamate
10	Sodium cocoyl glutamate
11	Sodium cocoyl aspartate
12	Sodium cocoyl methyl alanine
13	Sodium laureth-4 carboxylate

propyl) hydroxyethyl chloride cellulose). It gives a favorable feeling and long retains the "smooth (lubricious)" touch during rinsing, which is favored particularly by professional beauticians. There are polyquaternium-10 products of various cellulose sizes and cationic degrees, which differ in feel of use and viscosity. The feel of use and viscosity also depend on anionic surfactant to combine (Table 5.5).

It is important to select polyquartenium-10 suitable for the anionic surfactant component. For example, polyquartenium-10 that is highly cationic and low viscous is suitable for sodium laureth sulfate. Those that are little cationic and highly viscous are appropriate for amino acid surfactants.

Table 5.3 Difference in sensory characteristics of shampoo by anionic surfactants

Anionic surfactant	Lathering, size of bubbles	At application	Rinsing	Towel drying	Overall rating
Sensory evaluation (1)					
1	Rather good	Flexible, lubricating, good texture	Moist but weak lubrication	Heavy lubrication. Slightly rough	B
2	Rather good	Rough, coarse	Bad lubrication. Coarse	Bad lubrication. Coarse and rough	CD
3	Weak	Flexible, heavy lubrication, rough	Bad lubrication. Rough	Slightly better than 2	CD
4	Weak, small	Flexible, not lubricating	Bad lubrication	Less lubrication than 1. Rough	BC
5	Rather good	Worse than 1 and better than 4. Weak lubrication	Bad lubrication	Worse than 1 and better than 4	B
6	Rather good, big	Similar to 1. Moisturizing	Moist and persistent	Moist, lubricating, good texture	A
7	Rather good	Flexible, lubricating	Bad lubrication	Bad lubrication. Rough	BC
8	Good, small	Similar to 7	Moist, persistent, and favorable texture	Rather heavy lubrication. Rather rough	AB
9	Rather good, big	Similar to 7	Very moist, lubricating, and favorable texture	Rather heavy lubrication. Rather rough	AB
10	100	Moist. Strong lubrication	Moist, persistent, and favorable texture	Moist, lubricating, good finger pass	A
11	Weak, small	Heavier lubrication than 7	Moist, persistent, and favorable texture	Rather heavy lubrication. Rather rough	AB
12	Rather good, big	Highly flexible. Lubricating	Moist but weak lubrication	Rather heavy lubrication. Rather rough	AB
13	Good, big	Flexible but heavy lubrication and rough	Moist but weak lubrication. Rather coarse	Heavy lubrication. Rough	CD
Sensory evaluation (2)					
1	Normal	Favorable lubrication	Quick rinsing. No sliminess. Persistent but bad texture. Normal softness		C
2	Weak, small	Stronger lubrication than 1. Elastic and soft	Better than 1. Soft		C

3	Weak, small	Similar lubrication to 2	Similar to 1 in rinsing performance and sliminess persistence. Softer than 1		CD
4	Bad, small	Strongest lubrication. Very soft	No persisting sliminess. Rough. Lubricating and very soft	Soft	A
5	Good, rather small	Lubrication	Similar to 4 in rinsing performance. Less rough than 4		AB
6	Bad, coarse	Not lubricating and not soft	Similar to 3		AB
7	Good, rather small	Intermediate of 4 and 5 in lubrication, texture, lathering and softness	Strongly persistent. Weak lubrication and softness	Soft	A
8	Bad, small	Lubricating and soft	Lubricating and soft	Springy. Adhesive	B
9	Normal	Normal softness and lubrication	Most persistent. Breaking and thick feel	Light. Silky	AB
10	Normal, small	Lubricating and softer than 9	Lubricating and soft. Favorable springiness	Good lubrication	AB
11	Bad, very small	Rather lubricating and less softer than 9	Soft but unfavorable persistence		BC
12	Very good, normal	Normal lubrication and softness	Less coated feeling than 11 and 8. Light		B
13	Good, small	Heavy lubrication. Highly soft	Sliminess disappear quickly. Bad lubrication. Dry and hard		B

Note: *A*: good texture; *AB*: better than B; *B*: normal; *BC*: worse than B; *C*: light texture; *CD*: better than D; *D*: bad texture
The test results showed that the viscosity and amine number of Polyquaternium-10 affected the sensory characteristics of shampoos

Table 5.4 Characteristics of amphoteric surfactants in shampoo formulations

		Viscosity	Lathering	Sensory
Imidazolinium betaine	Sodium lauroamphoacetate	Normal	Rather low	Moist
Betaine	Cocamidopropyl betaine	Rather high	Good	Lubricating
Betaine	Lauryl betaine	High	High	Light
Amine oxide	Lauryl dimethyl amine oxide	Normal	Good	Rather moist

5.2.4.1 Merquat

Copolymers called "Merquat" are also widely used in shampoos, hair conditioners, and rinse-off products to improve the feel of use (See Sect. 2.3.2.2.1, Chap. 2, for the constituents of Merquat.).

The stability and feel of use of Merquat combined in transparent shampoos are shown in Table 5.6. Merquat improves the feel of use when combined with polyquaternium, and the feel differs depending on Merquat product.

5.2.4.2 Substitutes for Polyquaternium-10

Continuous use of polyquaternium-10 may cause complexes with anionic surfactants to remain on the hair, resulting in coarse feeling. Substitutes for polyquaternium-10 have been researched and developed.

Today, polyquaternium-7, polyquaternium-22, polyquaternium-39, polyquaternium-47, polyquaternium-52, polyquaternium-67, *Trigonella foenum*-gracecum hydroxypropyl trimonium chloride and guar hydroxypropyl trimonium chloride are widely used combined with other polymers. Guar hydroxypropyl trimonium chloride is also used alone in some products.

5.2.5 Pearlizing Agents

Glycol distearate and polyethylene glycol distearate are used as pearlizing agents either alone or combined with cetanol or oleyl alcohol. Pearl shampoos are prone to separation and viscosity drops and are difficult to maintain stable. Measures should be implemented to increase the viscosity and ensure stability, such as adding salt and selecting appropriate cationic polymers and anionic surfactants. Constituents that are insoluble in water and oils as well as those that make transparent shampoos turbid can be included into pearl shampoos. Ingredients for giving concepts and improving the feel of use can also be added. For example, jojoba oil, which is vegetable oil, and jojoba wax PEG-esters can be added to constitute about 1.5%.

Care should be taken in adding pearlizing agents because polyethylene glycol distearate melts only at high temperatures. Be sure to dissolve the agents thoroughly by heating to 70°C or above, stir well with other surfactants, and cool. Otherwise, undissolved portion educes, precipitates, and separates.

Table 5.5 Effects of anionic surfactants and polyquaternium-10 on sensory characteristics of shampoo

	1	2	3	4	5	6
	Sodium cocoyl methyl taurate	TEA-Cocoyl glutamate	Sodium PEG lauryl ether sulfate	Sodium cocoyl methyl taurate	TEA-Cocoyl glutamate	Sodium PEG lauryl ether sulfate
Anionic surfactant	36.00			36.00		
Cocamide propyl betaine	8.00	8.00	8.00	8.00	8.00	8.00
Cocamide DEA	4.00	4.00	4.00	4.00	4.00	4.00
Methyl paraben	0.20	0.20	0.20	0.20	0.20	0.20
Polyquaternium-10	(High viscosity, low amine value) 1.00			(Low viscosity, high amine value) 1.00		
50% citric acid	00.30	0.30	0.30	0.30	0.30	0.30
Sodium benzoate	0.20	0.20	0.20	0.20	0.20	0.20
Purified water	to 100 g	to 100 g	to 100 g	to 100 g	to 100 g	to 100 g
Lathering	Good lathering. Fine bubbles	Rather good lathering. Coarse bubbles	Good lathering. Coarse bubbles	Good lathering. Coarser bubbles than 1	Similar lathering to 2. Slightly coarser bubbles than 2	Similar lathering and bubble sizes to 1
Rinsing	Appropriate softness. Slimy lubrication	Less softness and lubrication than 1	Appropriate lubrication. Rather hard	Harder and less lubrication than 1	Harder and less lubrication than 2	Harder and less lubrication than 3
Towel drying	Persistent sliminess. Good lubrication. Appropriate softness	More persistent sliminess than 1. Less lubrication than 1	Similar sliminess persistent to 1. Less lubrication and softness than 1	More persistent sliminess than 1. Similar lubrication to 1	More persistent sliminess, lubrication and softness than 2	Similar softness to 3. Less persistent sliminess and lubrication than 3
Overall rating	AB	B	B	BC	C	C

Note: *A*: good texture; *AB*: better than B; *B*: normal; *BC*: worse than B; *C*: light texture; *CD*: better than D; *D*: bad texture
The test results showed that the viscosity and amine number of Polyquaternium-10 affected the sensory characteristics of shampoos

Table 5.6 Sensory characteristics of shampoos by Merquat

Basic prescription		
Polyquaternium Merquat		0.40
EDTA-2Na		0.10
Methyl paraben		0.10
Sodium benzoate		0.10
TEA-Laureth sulfate		28.00
Cocamidopropyl betaine		8.00
Cocamide DEA		3.00
50% citric acid		For adjusting pH
Purified water		to 100 g
Stability of Merquat		
	Stability (at room temperature)	Stability (refrigerated)
Merquat 100	Precipitate	
Merquat 280	Transparent	Transparent
Merquat 295	Precipitate	
Merquat 550	Transparent	Transparent
Merquat Plus 3330	Slightly clouding	Transparent
Merquat Plus 3331	Transparent	Transparent
Merquat 2001	Slightly clouding	Slightly clouding
Merquat 2003	Clouding	

Sensory evaluation					Damaged hair used	
Ingredient investigated	No Merquat	Merquat 280	Merquat 550	Merquat Plus 3330	Merquat Plus 3331	Merquat 2001
Amount		0.20%	0.80%	0.80%	0.80%	0.40%
At application	Low sliminess and lubrication. Slightly coarse	Low sliminess and lubrication. Not coarse	Low sliminess and lubrication	Slimy, lubricating, moist, and good texture	Less lubrication and sliminess. Moist and good texture	Slimy, lubricating, moist and good texture
During rinsing	Slightly lubricating and slightly coarse	Slightly lubricating and very slightly coarse	Rather heavy lubrication	Soft, moist, and good texture	Soft, moist, and good texture	Moist, persistent, and good texture
After rinsing	Heavy lubrication. Coarse	Heavy lubrication	Good lubrication	Smooth and lubricating	Smooth and light lubrication	Good lubrication

5.2.6 Influence of Salts

Salt affects the viscosity and low-temperature stability of products. Salts, which are electrolytes, cause salting-out of ionic surfactants, resulting in increased viscosity. In case of transparent shampoo, which uses POE alkylphenyl ether sulfate, the viscosity and low-temperature stability of the product increase as the salt concentration increases. However, in shampoo products that use triethanolamine-cocoyl glutamate or sodium menthyl cocoyl taurate, salts do not increase the viscosity but make the products prone to clouding and separation at low temperatures. In pearl shampoos, salt prevents separation and improves stability because it increases viscosity. In shampoos that use sodium lauryl sulfate, salts increase viscosity and thus stabilize the products. Products that use carboxyl surfactants show drops in viscosity and are thus difficult to maintain stable.

5.2.7 Thickeners and Bubble Stabilizers

Widely used thickeners and bubble stabilizers in shampoos are nonionic alkanolamide surfactants. Lauramide DEA and cocamide DEA are effective and most widely used in transparent shampoos. Cocamide MEA and cocamide MIPA are widely used in pearl shampoos but not in transparent shampoos because they may cause precipitation. Other thickeners and bubble stabilizers include glucoside surfactants such as PEG-120 methyl glucose dioleate and PEG-120 methyl glucose triisostearate. They increase the viscosity with a small amount but give a heavy feeling. Butylene glycol laurate is effective for anionic laureth sulfate surfactants but not for anionic carbonate surfactants. Alkylglucosides such as lauryl glucoside need to be included in a relatively large quantity to increase the viscosity. PEG stearate can thicken shampoos with a small quantity but makes heavy shampoos that take time to lather.

Agents for reducing the viscosity of shampoos include glycols and ethanol. Particularly 1,3-butylene glycol and propylene glycol are effective, and addition of several percent reduces the viscosity sharply.

5.2.8 Ingredients for Improving the Feel of Use

Ingredients used to improve the feel of use of shampoos include PEO vegetable oils, acylated hydrolyzed proteins, amino acids, silicone emulsions, amino-modified silicones, POE and POP silicones. Silicones are most widely used to enhance the silky feeling. Nonionic surfactants such as POE vegetable oil augment the moisturized feeling. Various PEO additives of vegetable oil origin are now available on the market. PEO lanolin was once used for professional use products.

Acylated hydrolyzed proteins are also effective for increasing the moisturized feeling. They are more effective than commonly used hydrolyzed proteins. The acylation also activates proteins.

5.2.9 *Efficacious Components and Conceptual Components*

Efficacious and conceptual components include hydrolyzed proteins, amino acids, ceramides, plant extracts, polysaccharides, and fruit acids. They must be combined so as to not form complexes with the other constituents or destabilize the product. Oils of high melting points may need solubilization. Components that are cationic or have amino groups may form complexes with the anionic surfactant component or cause separation of the product. Stability needs to be examined even for adding a very small amount.

5.2.10 *Stabilizers, Preservatives, and Acidity Regulators*

Stabilizers are generally added to block metal ions and should be selected based on the selectivity of metals. Use of two or more preservatives is more effective than using only one preservative. Challenge tests, which involve growing bacteria, should be performed because preservatives may lose efficacy by combinations of ingredients, pH, etc.

The pH of shampoos affects the viscosity and stability of the products. At low pHs, the viscosity increases and the product is prone to clouding at low temperatures.

The pH may rise by amines contained in alkanolamides. It is recommended to use citric acid, lactic acid, glycolic acid, or malic acid to adjust the pH. Lowering the pH of transparent shampoos may result in clouding and separation at low temperatures. Particularly those that have carbonate surfactants as the anionic surfactant component are prone to clouding and should be adjusted at pH 6 or above.

Shampoos may undergo coloring, smelling and changes in viscosity by temperature changes, and therefore must be thoroughly tested. Pearl shampoos are particularly prone to separation.

5.2.11 *Sensory Evaluation of Shampoo*

Sensory evaluation scores of shampoo vary greatly between home users and professional beauticians. The difference should be noted when deciding formulations. When home users shampoo the hair, they keep rinsing until the slippery feeling of shampoo disappears believing the shampoo is still remaining. They believe shampoos that do not rinse quickly are bad shampoos. On the other hand, beauticians think that shampoos that keep the hair smooth throughout the rinsing process are good shampoos.

Sensory evaluation involves taking a small quantity of shampoo on the palm, lathering, washing the hair, and rinsing away. Each process is evaluated. At the time of the application, shampoo is evaluated for lathering, the hardness of the foam, and whether the foam makes the hair feel soft and smooth or not. At time of rinsing, it is assessed whether the smoothness last long or disappears quickly, and for the stickiness to the hand, smoothness, and softness. Prior to conducting sensory evaluation, it is recommended to summarize the detailed check items in a table. The shampoo should then be evaluated for the feeling while toweling the hair: whether the hair retains the smooth feeling or the fingers tend to get caught in the hair. The scores of the evaluation reflect whether the combination of the anionic surfactants and the cationic polymers was appropriate for the purpose or not.

5.2.12 Methods of Adding Scenting Agents

To add scenting agents and essential oils into shampoos, it is recommended to mix the scenting agent into a solubilizing agent in advance. PEG-20 sorbitan cocoate and PEG-20 sorbitan oleate are widely used solubilizing agents. Mixing scenting and solubilizing agents at a ratio of 1:1–1:4 facilitates adding the scenting agent into transparent shampoos. Nonionic surfactants such as POE alkyl ethers and POE fatty acids can also be used as solubilizing agents, but it takes time to dissolve the mixture in shampoos because the surfactants hydrate with water and form gels. Dissolution can be accelerated by using the nonionic surfactants together with 1,3-butyl glycol or propylene glycol, but this may lower the viscosity of the product.

5.2.13 Designing Shampoo Formulations

In general, a shampoo consists of 30–40% anionic surfactants (assumed purity: 30%), about 10% amphoteric surfactants, 3–6% thickening agents such as alkanolamide and other nonionic surfactants in most cases, and 0.3–1.0% cationic polymers to improve the feel of use and increase viscosity. The sensory characteristics of shampoos are determined by the ratio between cationic polymers and anionic or amphoteric surfactants; and viscosity is also determined by the combination of these four ingredients. Stabilizers, preservatives, and acidity regulators are also important ingredients and should be thoroughly examined. Viscosity changes greatly by pH, and the changes are more conspicuous at lower pH. This is likely because the counter ions in anionic surfactants neutralize at low pH.

5.2.13.1 Cleansing Shampoo

Prescription 5.8 is a shampoo for removing hair oil, etc. and uses sodium methyl cocoyl taurate. It is also suitable for scalp cleansing shampoos.

The combination of sodium methyl cocoyl taurate and sodium C14–16 olefin sulfonate gives good lathering and a smooth texture. AMP and basic amino acids are added to help removing filth, and amino acids improve the feel of use.

Prescription 5.8 Cleansing shampoo

	Part	Ingredient	% (100 g)
1	A	Sodium methyl cocoyl taurate (30%)	18.00
2	A	Sodium C14–16 olefin sulfonate (37%)	9.00
3	A	Cocamide propyl betaine (30%)	8.00
4	B	Cocamide methyl MEA	3.00
5	B	Polyquaternium-10 (high viscosity, low amine)	0.50
6	B	EDTA-2Na	0.20
7	B	Sodium benzoate	0.20
8	C	PEG-20 sorbitan cocoate	0.90
9	C	Scenting agent	0.20
10	D	Methyl isothiazolione (10%)	0.10
11	D	AMP	0.30
12	D	Arginine	0.20
13	D	Glutamic acid	0.10
14	pH	50% citric acid	0.60
		Purified water	to 100 g

Directions
1) Disperse 5 in purified water, dissolve by heating to 80°C, and add 6 and 7. (B)
2) Mix 8 and 9, and homogenize. (C)
3) Add 1–4 to B. (A+B)
4) Cool to 45°C, and add C and 10–13. (B+A+C+D)
5) Adjust the pH with 14 and add water

5.2.13.2 Soft and Moisturizing Shampoo

Prescription 5.9 also uses sodium methyl cocoyl taurate. The shampoo softens and moisturizes hard hair. The combinations of the anionic surfactants of sodium lauroyl beta-alanine, sodium methyl cocoyl taurate, and sodium lauroyl hydrolyzed silk and the cationic polymers of cationated guar and polyquaternium-7 gives a soft touch.

Prescription 5.9 Soft and moisturizing shampoo

	Part	Ingredient	% (100 g)
1	A	Sodium lauroyl beta-alanine (30%)	27.00
2	A	Sodium methyl cocoyl taurate (30%)	8.00
3	A	Cocamide propyl betaine (30%)	9.00
4	A	Cocamide methyl MEA	4.00
5	B	Guar hydroxypropyl trimonium chloride	0.80
6	B	EDTA-2Na	0.20
7	B	Sodium benzoate	0.20
8	B	Sodium citrate	0.40
9	B	Betaine	0.50
10	B2	Polyquaternium-7	1.00
11	C	PEG-20 sorbitan cocoate	1.20
12	C	Scenting agent	0.20
13	C	Methyl isothiazolione (10%)	0.10
14	C	Sodium lauroyl hydrolyzed silk (20%)	2.00
15	C	Hydrolyzed collagen (20%)	0.50
16	pH	50% Citric acid	0.60
		Purified water	to 100 g

Directions

1) Disperse 5 in purified water, dissolve by heating to 80°C, add 6–9 and dissolve, and add 10. (B)
2) Mix 11 and 12, and homogenize. (C)
3) Add 1–4 to B. (B + A)
4) Cool to 45°C, and add C and 13–15. (B + A + C + D)
5) Adjust the pH with 16 and add water

5.2.13.3 Damage Care Shampoo

Prescription 5.10 combines anionic surfactants of sodium lauroyl beta-alanine and sodium methyl cocoyl taurate, cationic polymers of polyquaternium-10 and polyquaternium-7, and amino-modified silicone. The complex components of the shampoo adhere to the damaged hair, which has turned anionic and hydrophilic, and give a smooth touch to the hair.

Prescription 5.10 Damage care (moisturizing) shampoo

	Part	Ingredient	% (100 g)
1	A	Sodium lauroyl beta-alanine (30%)	27.00
2	A	Sodium methyl cocoyl taurate (30%)	9.00
3	A	Cocamide propyl betaine (30%)	9.00
4		Lauramidopropylamine oxide (30%)	3.00
5	A	PEG-120 methyl glucose dioleate	1.00
6	B	Polyquaternium-10 (high viscosity, low amine)	0.40
7	B	EDTA-2Na	0.10

(continued)

Prescription 5.10 (continued)

	Part	Ingredient	% (100 g)
8	B	Sodium benzoate	0.20
9	B2	Polyquaternium-7	1.20
10	C	PEG-20 sorbitan cocoate	0.60
11	C	Scenting agent	0.20
12	D	Methyl isothiazolione (10%)	0.10
13	D	Hydrolyzed keratin	1.00
14	D	Bis (C13-15 alkoxy) PG amodimethicone	0.60
15	pH	50% Citric acid	0.30
		Purified water	to 100 g

Directions
1) Disperse 6 in purified water, dissolve by heating to 80°C, add 7 and 8 and dissolve. Add 9. (B)
2) Mix 10 and 11, and homogenize. (C)
3) Add 1–5 to B. (B + A)
4) Cool to 45°C, and add C and 12–14. (B + A + C + D)
5) Adjust the pH with 15 and add water

5.2.13.4 Silky Shampoo (Also for Hair Extensions)

Prescription 5.11 is suitable for rough hair and can also be used for hair extensions. Sodium laureth sulfate is combined with disodium laureth sulfosuccinate to enhance lubrication, and amino-modified silicone is added to improve smoothness.

Prescription 5.11 Damage care shampoo

	Part	Ingredient	% (100 g)
1	A1	Sodium laureth sulfate (30%)	28.00
2	A1	Cocamide DEA	3.00
3	A	Cocamide propyl betaine (30%)	9.00
4	A	Disodium laureth sulfosuccinate (30%)	8.00
5	B	Polyquaternium-10 (high viscosity, low amine)	0.60
6	B	EDTA-2Na	0.10
7	B	Sodium benzoate	0.20
8	B	Sodium chloride	0.30
9	C	PEG-20 sorbitan cocoate	0.60
10	C	Scenting agent	0.20
11	D	Methyl isothiazolione (10%)	0.10
12	D	Bis (C13-15 alkoxy) PG amodimethicone	0.50
13	pH	50% Citric acid	0.30
		Purified water	to 100 g

Directions
1) Heat purified water to 80°C, and dissolve 5–8. (B)
2) Mix 9 and 10, and homogenize. (C)
3) Add 1–4 to B. (B + A)
4) Cool to 45°C, and add C, 11 and 12. (B + A + C + D)
5) Adjust the pH with 13 and add water

5.2.13.5 Volume Up Shampoo

Prescription 5.12 is for adding springiness to thin and soft hair. Sodium cocoamphoacetate is used instead of cocamide propyl betaine, and lauryl trimonium chloride is added to form complexes with anions and give persistent and adhesive feeling. Stearyl trimonium chloride gives a stronger adhesive feeling and less lubrication than lauryl trimonium chloride.

Prescription 5.12 Volume up shampoo

	Part	Ingredient	% (100 g)
1	A	Sodium laureth sulfate (30%)	36.00
2	A	Sodium cocoamphoacetate (30%)	8.00
3	A	Lauryl trimonium chloride (30%)	0.50
4	A	Cocamide DEA	3.00
5	B	Polyquaternium-10 (high viscosity, low amine)	0.50
6	B	EDTA-2Na	0.10
7	B	Sodium benzoate	0.20
8	B	Citric acid	0.20
9	C	PEG-20 sorbitan cocoate	0.60
10	C	Scenting agent	0.20
11	D	Methyl isothiazolione (10%)	0.10
12	pH	50% Citric acid	
		Purified water	to 100 g

Directions
1) Disperse 5 in purified water, dissolve by heating to 80°C, and dissolve 6–8. (B)
2) Mix 9 and 10, and homogenize. (C)
3) Add 1–4 to B. (B + A)
4) Cool to 45°C, and add C and 11. (B + A + C + D)
5) Adjust the pH with 12 and add water

5.2.13.6 Menthol Shampoo

Prescription 5.13 contains menthol and is an especially popular shampoo in summer. Menthol is solubilized with polybutylene glycol 3PEG/PPG-8/5 glycerin, ethanol, and PEG-20 sorbitan cocoate to produce a transparent shampoo.

Prescription 5.13 Menthol shampoo

	Part	Ingredient	% (100 g)
1	A	Polyquaternium-10 (high viscosity, high amine)	0.50
2	A	EDTA-2Na	0.10
3	A	Citric acid	0.10
4	B	Sodium laureth sulfate (30%)	36.00
5	B	Cocamide propyl betaine (30%)	9.00
6	B	Cocamide DEA	4.00

(continued)

Prescription 5.13 (continued)

	Part	Ingredient	% (100 g)
7	C	Methyl paraben	0.20
8	C	Propyl paraben	0.10
9	C	Salicylic acid	0.20
10	C	Polybutylene glycol 3PEG/PPG-8/5 glycerin	4.00
11	C	Ethanol	2.00
12	C	PEG-20 sorbitan cocoate	4.00
13	C	Menthol	4.00
14	D	Bis (C13-15 alkoxy) PG amodimethicone	0.50
15	pH	50% Citric acid	0.20
		Purified water	to 100 g
		pH	5.80

Directions
1) Disperse 1 in purified water, and dissolve by heating to 70°C. Add 2 and 3, and dissolve. (A)
2) Add 4–6 to A. (A + B)
3) Dissolve 7–9 in 10 and 11, and dissolve 12 and 13. (C)
4) Cool A and B to 45°C, and add C. (A + B + D)
5) Add 14. (A + B + C + D)
Adjust the pH with 16 and add water.

Specifications
pH: 5.0–6.0

5.2.13.7 Natural Oil Shampoo

Prescription 5.14 consists of natural oils, natural polyoxyethylene oil, and isostearyl alcohol as a solubilizing agent. It gives a moisturized feeling and makes the hair that is easy to unite.

Prescription 5.14 Oil-rich conditioning shampoo (non-silicone)

	Part	Ingredient	% (100 g)	% (100 g)	% (100 g)	% (100 g)
1	A	Polyquaternium-10 (high viscosity, high amine)	0.50	0.50	0.50	0.50
2	A	EDTA-2Na	0.10	0.10	0.10	0.10
3	A	Sodium chloride	0.80	0.80	0.80	0.80
4	A	Sodium benzoate	0.20	0.20	0.20	0.20
5	A	Methyl paraben	0.20	0.20	0.20	0.20
6	A	Propyl paraben	0.10	0.10	0.10	0.10
7	B	Sodium laureth sulfate (30%)	36.00	36.00	36.00	36.00
8	B	Cocamide propyl betaine (30%)	9.00	9.00	9.00	9.00
9	B	Cocamide DEA	3.00	3.00	3.00	3.00
10	B	Ethylene glycol distearate	2.50	2.50	2.50	2.50

(continued)

Prescription 5.14 (continued)

	Part	Ingredient	% (100 g)	% (100 g)	% (100 g)	% (100 g)
11	B	Jojoba oil		0.50	0.50	0.50
12	B	Isostearyl alcohol		0.10	0.10	0.10
13	B	Jojoba wax PEG-80			1.50	
14	B	Jojoba wax PEG-120				1.50
15	pH	50% Citric acid	0.40	0.40	0.40	0.40
		Purified water	to 100 g	to 100 g	to 100 g	to 100 g
		pH	4.20	4.50	4.50	4.50
Feel of use		Lathering during application	Good	Good	Good	Good
		Rinsing (combing 10 times under running water	Rough	Slightly rough but lubricating	Lubricious and good	Lubricious and smooth
		Towel drying	Not very lubricious and light	Not very lubricious and light	Lubricious and good	Lubricious and good

Directions
1) Disperse 1 in purified water, and dissolve by heating to 75°C. Add 2–6, and dissolve. (A)
2) Add 7–14 to A, stir, and cool to 45°C. (B)
3) Adjust the pH with 15 and add water

Specifications
pH: 4.0–6.0

5.2.13.8 Transparent Zein Shampoo

Prescription 5.15 is adhesive and coats the hair surface. This transparent shampoo contains zein, which is insoluble in water and oil. Zein becomes soluble in water by combining with DPG. To combine in shampoos, coagulation can be minimized by diluting with PEG-200, PEG-400, PG, and/or DPG, and a transparent solution can be obtained.

Prescription 5.15 Zein shampoo

	Part	Ingredient	% (100 g)
1	A	Polyquaternium-10 (high viscosity, high amine)	0.60
2	A	EDTA-2Na	0.10
3	A	Betaine	0.30
4	A	Methyl paraben	0.20
5	A	Propyl paraben	0.10
6	B	Sodium methyl cocoyl taurate (30%)	9.00
7	B	TEA-cocoyl glutamate	27.00
8	B	Cocamide propyl betaine (30%)	9.00
9	B	Lauramidopropyl dimethylamine oxide	3.60
10	B	PPG-2 cocamide	2.40
11	C	10% sodium hydroxide	0.50

(continued)

Prescription 5.15 (continued)

	Part	Ingredient	% (100 g)
12	D	Zein, DPG (*2)	0.50
13	D	PEG-200	0.50
14	pH	50% Citric acid	
		Purified water	to 100 g

Directions
1) Disperse 1–5 in purified water, dissolve by heating to 75°C to make a transparent solution. (A)
2) Add 6–10 to A and stir. (A+B)
3) Cool to 60°C, and add 11. (A+B+C)
4) Mix 12 and 13, and add to (A+B+C). (A+B+C+D)
5) Adjust the pH with 14 and add water

Specifications
pH: 5.8–6.5
*2 Phytokeluster Z (Ichimaru pharcos Co., Ltd.)

5.3 Hair Conditioners

Hair conditioners are rinse-off cosmetics that are usually used after shampooing hair. They include products called "hair rinse," "hair conditioner," "hair treatment," and "hair pack." The method of use differs slightly by the name. Hair rinses and conditioners are always used after shampoos to reduce the roughness caused by the shampoo. Hair treatments and packs are deemed as products to repair hair damages.

In Japan, the term "hair rinse" is most widely used because it is used in the administrative procedure for manufacturing and selling cosmetics products. However, in this book, they are referred to as "hair conditioners" or simply as "conditioners." There are little differences in constituents among hair rinses, conditioners, treatments and packs, but hair treatments and hair packs also contain proteins, amino acids, moisturizing ingredients, and/or other efficacious constituents in general to repair damage.

5.3.1 Sensory Characteristics of Conditioners

Conditioners are used to improve the feel of the hair after shampooing so as to be smooth and moist. Users select the feel given by conditioners depending on the characteristics of the hair, personal preference, and concepts of the product. Therefore, the sensory characteristics must be designed so as to suit the purpose of use. Unlike in shampoos, the demanded sensory characteristics seldom vary between home users and professional beauticians.

The sensory characteristics of conditioners are determined mainly by the cationic surfactants used. Higher alcohols also affect the sensory characteristics and viscosity. Silicones increase the lubricious, smooth, and silky feeling. Vegetable oils, esters, and carbohydrates also affect the feel of use but to a lesser degree.

Conditioners may also contain nonionic surfactants, glycol and polymers to improve emulsification and stability. They also affect the sensory characteristics, and sometimes are used to achieve the aimed feel. Conceptual and efficacious ingredients, preservatives, antioxidants, acidity regulators, and scenting agents are also indispensable.

5.3.2 Adjusting the Viscosity of Conditioners

The feel of use and viscosity should be determined simultaneously based on the purpose and container of the product. The viscosity of conditioners is roughly determined by the amounts of higher alcohols and cationic surfactants. It is easy to start by investigating the standard contents of 3% cetrimonium chloride (70%) and 6% cetanol. Increasing the amount of cetanol results in increased viscosity, and decreasing the amount lowers the viscosity. The viscosity should be adjusted by not altering the sensory characteristics as much as possible because the feel of use is crucial in conditioners. Therefore, it is better to change the alkyl group, polarity and melting point of oily constituents rather than changing the cationic surfactant.

Chronological viscosity changes and viscosity changes by temperature should be examined (See Sects. 3.1.3.1 and 3.2 in Chap. 3 for methods of adjusting viscosity.) (Fig. 5.1).

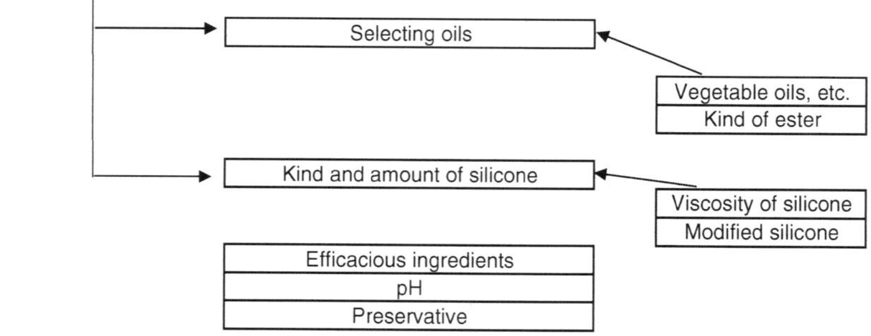

Fig. 5.1 Process of investigating formulations of hair conditioners

5.3.3 Basic Prescriptions and Constituents

The basic prescriptions and typical constituents of hair conditioners are shown in Table 5.7. The sensory characteristics can be easily changed by changing cationic surfactants and their combinations, higher alcohols and their combination, and silicones and their combinations.

Table 5.7 Common formulation of hair conditioner

Component	Amount	Widely used ingredients	
Cationic surfactants	1–4%	Behentrimonium chloride	Emulsifier
		Steartrimonium chloride	Best emulsifier
		Cetrimonium chloride	Emulsifier
	0–2%	Methyl sulfate salt	Emulsifier
	0–2%	Dialkyl dimonium chloride	Feel improver, emulsifier
Higher alcohols	3–8%	Cetanol	Base, thickener, feel improver
		Cetostearyl alcohol	
		Stearyl alcohol	
	0–1%	Octyldodecanol, Isostearyl alcohol	Viscosity adjuster, oil
	0–1%	Oleyl alcohol	Oil
Esters	0–3%	Cetyl ethylhexanoate	Feel improver
		Ethylhexyl palmitate	
		Hexyldecyl isostearate	
		Isopropyl palmitate	
		Isopropyl myristate	
Hydrocarbons	0–6%	Squalane	Add texture
		Liquid paraffin	
Nonionic surfactants	0–1%	Polyoxyethylene cetyl ether	Emulsification aid, add when necessary
		Glyceryl monostearate	
Silicones	1–10%	Decamethylcyclopentasiloxane	Lubrication, smoothness
		Methylpolysiloxane polymer	
		Methylpolysiloxane	
	0–2%	Stearoxy dimethicone	
		Aminoethylaminopropyl dimethicone	
		Methylpolysiloxane copolymer	
Vegetable oils	0–6%	Jojoba oil	Moisturizing, softness
		Macadamia nut oil	
		Olive oil	
Viscous oils	0–6%	Lanolin and similar oils	Moisturizing
		Dimer dilinoleyl esters	Adhesion
		Pentaerythrityl fatty acid	Persistence
Glycols	0–4%	Glycerin, dipropylene glycol	
Polymer	0–0.5%	Hydroxy ethyl cellulose	Stabilizer

(continued)

Table 5.7 (continued)

Component	Amount	Widely used ingredients
Antioxidants	0–0.1%	Natural Vitamin E, Tocopheryl acetate, BHT
Preservatives	<1%	Methyl paraben, propyl paraben
Acidity regulators		Lactate, citrate, glycolate, and their salts
Scenting agents		Perfume, essential oils
Coloring agents		Tar dyes, plant pigments, caramel, licorice extract
Efficacious ingredients	Small amount	Hydrolyzed proteins, amino acids
	Minute amount	Ceramides, sterols
	Minute amount	Plant extracts, ingredients of plant origin
	Minute amount	Natural ingredients

5.3.3.1 Cationic Surfactants

The main cationic surfactant in conditioners is either cetrimonium chloride, steartrimonium chloride, or behetrimonium chloride, which have high emulsification capacities and sensory performances. Instead of monoalkyl cationic surfactants of C16, C18, and C22, dialkyl surfactants may be used such as dicocoyl dimonium chloride (there are also C12–14), dicetyl trimonium chloride and distearyl trimonium chloride but they should be combined with a monoalkyl surfactant to compensate for the low emulsification capacity.

Couple ions can be chlorides, bromides, and methyl sulfates. Chlorides have stronger emulsification capacity than bromides and methyl sulfates. The emulsification capacity, feel of use, and viscosity of creams change by the alkyl group of the cationic surfactant.

When the same prescription is emulsified, emulsification cannot be achieved by C12 lauryl. Cetyl (C16), which has high emulsification capacity, results in relatively low viscous cream that gives a light texture. Behenyl (C22) is a weak emulsifier and cannot produce cream by itself but needs to be combined with nonionic surfactant of low HLB. It produces viscous cream that gives a moist and flexible texture. Stearyl (C18) is intermediate of cetyl and behenyl and is the most widely used cationic surfactant of high emulsification capacity. It produces creams of favorable viscosity, which gives a lubricious and smooth texture (Table 5.8).

Table 5.8 Combinations of cationic surfactants and sensory characteristics of conditioners

Ratio		Feel of use	
2 to 4 parts	1 part	Strong sense	
Cetrimonium chloride	Dicocoyl dimonium chloride	Lubricating	Silky
Steartrimonium chloride	Dicocoyl dimonium chloride	Lubricating	Silky
Cetrimonium chloride	Steartrimonium chloride	Moist	Lubricating
Steartrimonium chloride	Cetrimonium chloride	Moist	Lubricating
Cetrimonium chloride	Behentrimonium chloride	Moist	Soft
Steartrimonium chloride	Behentrimonium chloride	Soft	Moist
Steartrimonium chloride	Distearyl dimonium chloride	Soft	Moist
Behentrimonium chloride	Distearyl dimonium chloride	Adhesive	Soft

5.3.3.2 Higher Alcohols

The number of carbons and structure (presence of double bonds and branch chains) of higher alcohol also affect the viscosity of creams. Of higher alcohols of C12, 14, 16, 20 and 22, C16 produces the most viscous cream from the same combination of ingredients (Fig. 5.2). The C12 alcohol is strongly hydrophilic due to the OH group at an end of the molecule. C22 alcohols, which have more carbon atoms than C12 alcohol, are less hydrophilic because the OH group has smaller effect. Cetanol, which has 16 carbon atoms, produces creams that are most viscous and affinitive to water. Oleyl alcohol produces low viscous creams. Branched-chain higher alcohols produce even less viscous creams. Other than cetanol and stearyl alcohol, combinations of myristyl alcohol and behenyl alcohol are promising candidates but are difficult to increase viscosity.

The length of the alkyl group and the presence of double bonds and/or branches also affect the feel of use. Compared to the most widely used cetanol, lauryl alcohol gives an inferior and rough texture, has a peculiar smell, and is thus rarely used.

To achieve favorable textures, it is effective to add a small amount of behenyl alcohol, oleyl alcohol, octyldodecanol, or isostearyl alcohol to cetanol. Behenyl alcohol enhances softness, oleyl alcohols gives a moist feeling, isostearyl alcohol results in light and lubricious texture, and octyldodecanol makes smooth and silky creams.

Higher alcohols of shorter chains result in lighter texture, and those that have more carbon atoms give moister feeling. Myristyl alcohol is unsuitable for rinse-off products because it gives a very light touch. Oleyl alcohol, which has a double bond, feels oily and heavy. Isostearyl alcohol and octyldodecanol are suitable for giving light and lubricious feeling; and the latter is especially suitable for rinse-off conditioners.

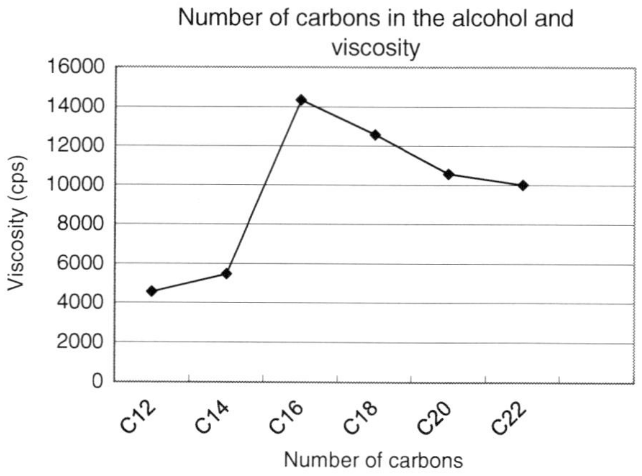

Fig. 5.2 Effects of higher alcohols on hair conditioner product

5.3.3.3 Silicones

Two or more silicones are frequently combined in hair conditioners, commonly in combination of highly polymerized dimethicone and an amino-modified silicone. The combination should be investigated for the viscosity of dimethicone and the amount of amino-modified silicone. Almost all hair treatment products contain a large amount of dimethicone. Solvents for highly polymerized dimethicone are also included, such as dimethicones of 10–200 cs, cyclopentasiloxane, and isododecane. The solvents affect the feel of use. Isododecane gives a heavy feel; and cyclopentasiloxane results in a light touch. Dimethicone is most suitable for rinse-off products. Water-soluble hydrophilic silicones, which have POE and/or POP, are unsuitable for improving sensory characteristics likely because the POE and POP groups are hydrophilic.

Amodimethicone is the most widely used amino-modified silicone. It has an aminopropyl group attached to the methyl group of dimethicone. Amodimethicone of various degrees of amino modification are available as well as those that have POP, POE, or an alkyl group attached. Amino-modified silicones are cationic and affinitive to hair keratin. They are particularly highly affinitive to damaged hair, which is anionic due to the presence of cysteic acid. Silicone content is several to 10% in general. An appropriate premix contains 2–6% of highly polymerized dimethicone, 2–3% of dimethicone of intermediate viscosity, and 1% or less of amino-modified silicone.

Method of Adding Silicones

The sensory characteristics of hair conditioner products are also affected by the timing of adding silicones. Generally, all oily ingredients are added together with surfactants into the oil phase and emulsified. However, in hair conditioners, silicones are added after the emulsification process at temperatures lower than 45°C so as to disperse in the aqueous phase. If silicones are added into the oily phase, they become thoroughly blended with the other oily ingredients and cannot fully manifest their sensory improvement effects. Silicones that are dispersed in the aqueous phase adhere easily to hair and improve the texture. Another reason for adding silicones after the emulsification process is because silicones have low compatibility with cationic surfactants and oily ingredients used in hair conditioners and are difficult to emulsify uniformly. It should be noted that cyclopentasiloxane is volatile and volatilizes as white gas at emulsification temperatures. Amino-modified silicones undergo chronological yellowing when exposed to high temperatures, which is a cause of coloring of creams.

It is difficult to disperse silicone oils in low viscous creams by stirring. Use of silicone emulsion is effective in such a case. Silicone emulsions, in which silicone is emulsified with cationic or nonionic surfactant, are easy to disperse but tend to give a light texture.

5.3.3.4 Vegetable and Animal Oils and Fats

Vegetable and animal oils and fats are mainly used as conceptual ingredients. They also give moist and smooth texture and may be included in 2–3% to improve the feel of use. Natural oils contain a large amount of unsaturated fatty acids, such as oleic acid, linolenic acid, and linoleic acid, which are easily oxidized, and are thus prone to smelling and color changes.

Oils and fats of relatively low iodine number should be used, and pH should be lowered to prevent oxidization. Use of antioxidants, such as tocopherol and lecithin, is also effective.

It is recommended to use oils of low iodine number such as those extracted from macadamia nut, olive, and meadowfoam. Shear butter, which is semi-solid, is also suitable.

5.3.3.5 Esters

Esters are combinations of fatty acids and alcohols, and are thus numerous. They have a polarity intermediate of carbohydrates and higher alcohols, and improve the compatibility of oily ingredients. They facilitate emulsification and produce creams that hardly undergo chronological viscosity changes.

Cetylethylhexanoate, cetylisostearate, and other liquid esters that have a branched-chain fatty acid or alcohol give lubricious and favorable textures. Solid esters that have straight-chain fatty acid and straight-chain alcohol, such as cetylpalmitate and stearyl stearate, give not so smooth but light and hard textures. Dimer acid esters such as dilinoleyl dilinoleate give an adhesive and persistent feeling. Esters that have a complicated structure, such as castor oil and pentaerythrityl fatty acids, give a moist and oily texture.

5.3.3.6 Waxes

Waxes have high melting points and long alkyl groups and easily adsorb to hair. Particularly, lanolin gives a favorable moist and smooth feeling to conditioners. Beeswax contains fatty acids, forms complexes with cationic surfactants, gives an adhesive and persistent texture, but reduces lubrication. Jojoba oil is liquid and gives a smooth texture. Candelilla wax and carnauba wax are solids of high melting points, give a hard and persistent texture, and are suitable for making the hair firm.

Adding liquid ester is effective for adding waxes into hair conditioners and emulsifying into fine-texture creams.

Of carbohydrates, liquid paraffin is relatively widely used. It gives a light and oily texture. Microcrystalline wax and paraffin wax can give a strong persistent texture by combining with candelilla or carnauba wax.

5.3.3.7 Polymers

Polymers are included into hair conditioners to achieve stabilization, thickening and sensory improvement. Hydroxyethyl cellulose and cationated guar gum are used to stabilize low viscous conditioners at high temperatures. However, addition of polymers into low viscous creams may result in destabilization, and therefore, thorough testing in advance is indispensable.

Merquats

Polyquaternium-10 and Merquat (Lubrizol Corporation) are effective for improving the sensory characteristics of rinse-off products such as hair conditioners.

The feel during application, rinsing and towel drying can be separately adjusted by combining with other cationic polymers. Merquat can improve the sensory characteristics of conditioners, multi-drug treatment products, hair dyes, and permanent wave agents during rinsing and the texture of the treated hair. There are Merquat products of various properties, which can be used to improve the feel of products by combining and forming complexes with anionic surfactant, cationic surfactant, other polymers, or another Merquat.

Sensory evaluation tests have been conducted on conditioners that contained tertiary amines and Merquat. For all tertiary amines, Merquat improved the coarseness and roughness. The tests also showed differences in texture by Merquat product (Tables 5.9 and 5.10).

5.3.3.8 Glycols

Glycerin and dipropylene glycol are used to stabilize creams, give luster, and improve spreadability. Propylene glycol and 1,3-butylene glycol are also used for similar purposes but may result in lowered viscosity and light texture. Polyethylene glycol, propylene glycol and its derivatives, and derivatives of glucosides can make the hair felt soft when applied.

5.3.3.9 Proteins and Amino Acids

Hair mainly consists of keratin and can be repaired from damage by supplying proteins. Proteins are included in hair conditioners mainly to prevent damages and repair the hair. The amount and kind of protein influence the feel of use of the product and the texture of the hair.

Widely used proteins include collagen, keratin, silk protein, and soybean protein. There are hydrolyzed products of these proteins of various molecular weights. There are also cationated, acylated, and silylated protein products.

Amino acids include neutral amino acids, such as glycine, basic amino acids such as arginine, and acidic amino acids such as glutamate and aspartic acid. Most amino acids are easy to combine into conditioners by dissolving in water. They have almost no influence on the sensory characteristics of conditioners products because the amount is small.

Points to Note for Adding Hydrolyzed Keratin

Hydrolyzed keratin of small molecular weight is easy to add into emulsions. It just involves adding the undiluted solution of hydrolyzed keratin into an emulsion at low temperature. On the other hand, hydrolyzed keratin of large molecular weight contains a large amount of glutamate and is thus anionic. It is vulnerable to reacting with cationic surfactants, forming complexes, destroying the emulsion, and resulting in a cream of coarse texture.

An effective and safe way to add a solution of a protein into an emulsion involves adding it into an aqueous solution of steartrimonium chloride in advance. This allows the proteins to react and form complexes with steartrimonium chloride, and makes the dispersion cationic, which is safely mixed into an emulsion.

Polyglutamic Acid and Polyaspartate

Polyglutamic acid and polyaspartate are polymers that have dense carboxyl groups due to the high acidity of the amino acids. Therefore, they make cationic surfactants to orient densely and strongly hydrophobic. Even at high cation concentration, complexes coagulate. Thus, polyglutamic acid and polyaspartate cannot be included in conditioners.

Table 5.9 Sensory evaluation of tertiary amines and polymers for improving the feel of use in hair conditioners

Basic formulation	A	B
Ingredient	% (100 g)	% (100 g)
Tertiary amine	Stearamido propyl dimethylamine 2.5	Behenamido propyl dimethylamine 2.5
Glyceryl monostearate	2.00	2.00
Behenyl alcohol	4.00	4.00
Myristyl alcohol	2.00	2.00
Dimer dilinoleate copolymer (phytostearyl, isostearyl, cetyl, stearyl, behenyl)	0.20	0.20
Rice bran oil	2.00	2.00
Cetyl ethylhexanoate	4.00	4.00
Polymers for improving the feel of use	n	n
Lactic acid (90%)	1.00	1.00
Purified water	to 100 g	to 100 g
pH: 3.0–5.0		

5.3.4 Sensory Evaluation

The sensory characteristics of conditioner products are evaluated by actually using the products. Hair conditioners are generally used after shampooing the hair by taking a small portion on the palm, applying on the wet hair, spreading, rinsing, wiping with a towel and drying with a blower. The senses while rinsing, towel drying and finishing (after drying) are evaluated separately by using hair tufts, wigs, and human hair. Once skilled, one can perform precise sensory evaluation.

Main sensory items to check for each process are shown in Table 5.11

Conditioners of various sensory characteristics can be designed by selecting the cationic surfactants to combine.

The moisturized feeling increases and the dry feeling decreases as the length of the alkyl group increases from laurtrimonium chloride to cetrimonium chloride, steartrimonium chloride, and behentrimonium chloride, in this order. Laurtrimonium chloride is unsuitable for hair conditioners. Dicetyl dimonium chloride and distearyl dimonium chloride give favorable texture when combined with monoalkyl surfactants. Differences by coupling ions slightly vary among chlorides, bromides, and methyl sulfates. Sensory characteristics by cationic surfactants are shown in Tables 5.8 and 5.12.

Table 5.10 Results of sensory evaluation

Polymer (pure fraction)	Amount (n)	At application	While rinsing	While towel drying	After drying
A) Tertiary amine: Stearamido propyl dimethylamine					
Hydroxyethyl cellulose	0.70	Good spread. Rather light texture	Smooth. United hair ends	Rather persistent. Rough middle section	Silky finish. Rather frizzy ends
Polyquaternium-10	0.30	Good spread but soft. Ordinary texture	Smooth. United hair ends	Rather hard texture. United hair ends	Hard texture. Hair ends slightly united
Merquat 199 (40%)	0.75	Good spread but soft. Ordinary texture	United hair. Good lubrication	Smooth and favorable texture. United hair	Moist texture but not heavy. Good
Merquat 280 (39%)	0.77	Good spread. Smooth texture	Soft texture	Moist and rather heavy texture	Moist texture but not heavy. Smooth and good
Merquat 295 (40%)	0.75	Spread well but require time to blend in hair	Moist and rather heavy texture	Heavy texture. Excessive unity	Springy and firm hair
Merquat 2001 (21%)	1.43	Spreads well but requires time to blend in hair	Moist texture. United hair ends	Moist texture. United hair ends	Rather too heavy
B) Tertiary amine: behenamido propyl dimethylamine					
Polymer (pure fraction)	Amount (n)	At application	While rinsing	While towel drying	After drying
Hydroxyethyl cellulose	0.70	Spreads well but requires time to blend in hair	Rather hard texture	Rather rough	Hard texture. Silky finish. Rough
Polyquaternium-10	0.30	Good spread. Moist texture	Moist. United hair ends	Rather coarse	Slightly rough
Merquat 199 (40%)	0.75	Good spread. Moist texture	Moist and rather heavy texture	Moist. United hair ends	Rather hard texture but smooth
Merquat 280 (39%)	0.77	Rather hard and difficult to spread	Rather hard texture	heavy texture. United hair ends	Hard and heavy. Good for thin hair
Merquat 295 (40%)	0.75	Rather hard and difficult to spread	Rather coarse texture	Rather rough	Good texture but rough hair ends
Merquat 2001 (21%)	1.43	Rather hard and difficult to spread	Heavy texture. Moist. Firmly united hair	Rather rough but united hair ends	Hard texture

Merquat improved the coarseness and roughness for both tertiary amines. The feel of use differed by the kind of Merquat used

The alkyl group determines the sensory characteristics of oily components. Oils that have larger molecular weight and longer alkyl group give stronger moisturized feeling. A double bond in the alkyl chain enhances the oily and moisturized touch. Those that have a branched structure tend to be silky and light. Oils that have complicated molecular structures give a moist and adhesive feeling. Because there are oils of various chemical structures, products of the aimed sensory characteristics are designed by investigating combinations of oils. The feels given by oils are summarized in Table 5.13.

Because feels and senses are somewhat personal, each product should be evaluated by at least two people and repetitively. Test hair specimens of uniform quality are difficult to obtain, and manifestation of sensory characteristics may differ depending on the degree of damage. In general, sensory characteristics are more apparent in more damaged hair specimens.

Senses felt may vary by person as well as the standards and expressions used. Sensory evaluation may also differ between home users and professional beauticians, and is sometimes entirely opposite. Analyzing the expressions used by each evaluator and his or her trends and summarizing the results as objectively as possible is a way toward correct evaluation.

Table 5.11 Sensory evaluation items for hair conditioners

Process	Sensory evaluation item	Content
At application	Blend into hair	Ease of blending the cream into hair
	Spread at application	Spread of the cream throughout the hair
	Sliminess, smoothness	Coarseness of the hair
	Softness, hardness	Enhanced softness by the cream
While rinsing	Retention of sliminess while rinsing	Long retention of smoothness while rinsing the hair
	Lubrication	Lubricated and smooth feeling
	Softness, hardness	Soft or hard
While towel drying	Lubrication	Lubricated and smooth
	Softness, hardness	Soft or hard
	Moistness, silkiness	Oily or not
	Coarseness, roughness	Rough and/or coarse feeling
	Adhesiveness, persistence, dryness	Oil coated feeling or not
Finished hair (after drying)	Lubrication	Lubricated and smooth feeling
	Softness, hardness	Soft or hard
	Moistness, silkiness	Moist or silky
	Adhesiveness, persistence, dryness	Oil coated feeling or not
	Uniformity	Same texture from the hair root to hair ends
	Unity	Whether hair tuft spreads or not

Table 5.12 Formulations of hair conditioners of various textures and sensory characteristics

	Coating	Moisturizing	Lubricating	Silky	Hard	Soft
Type of conditioner	Smooth and persistent	Moist and smooth	Lubricating and smooth	Light and smooth	Lubricating and making the hair springy	Smooth and soft
Cetrimonium chloride (70%)	–	–	–	–	1.20	1.20
Steartrimonium chloride (70%)	0.80	–	1.80	1.80	1.80	–
Behetrimonium chloride (80%)	1.60	2.80	–	1.20	–	1.80
Dicocotrimonium chloride (75%)	–	–	0.80	–	–	–
Disteartrimonium chloride (75%)	0.90	–	–	–	–	–
Glyceryl stearate	1.20	1.20	1.20	–	–	1.20
Cetanol	4.80	4.20	5.60	4.80	4.80	4.60
Behenyl alcohol	–	0.60	–	–	–	0.60
Oleyl alcohol	–	0.60	–	–	–	–
Octyldodecanol	–	–	0.60	0.60	–	–
PEG-30 cetyl ether	–	–	–	–	0.40	–
Microcrystalline wax	2.40	–	–	–	2.40	–
Paraffin	–	–	–	1.80	–	–
Cetyl ethylhexanoate	0.80	–	1.20	1.20	1.20	–
Castor oil	–	1.20	–	–	–	–
Olive oil	–	–	1.20	–	–	1.80
Macadamia nut oil	–	1.20	–	–	–	1.80
Dimer dilinoleyl dimer dilinoleate	–	0.40	–	–	–	–
Dimer dilinoleyl dimer dilinoleate glyceryl tri-hydrogenated rosinate	–	–	–	–	0.80	–
Propyl paraben	0.10	0.10	0.10	0.10	0.10	0.10
Methyl paraben	0.10	0.10	0.10	0.10	0.10	0.10
Aminoethyl aminopropyl dimethicone copolymer	–	0.80	0.80	–	–	0.80

(continued)

Table 5.12 (continued)

		Coating	Moisturizing	Lubricating	Silky	Hard	Soft
Type of conditioner		Smooth and persistent	Moist and smooth	Lubricating and smooth	Light and smooth	Lubricating and making the hair springy	Smooth and soft
Dimethicone 10cs, dimethicone (highly polymerized, 10%)		2.00	2.00	4.00	4.00	3.00	3.00
Dimethicone 1000cs		1.00	1.00	1.00	1.00	1.00	1.00
Purified water							
At application	Blend	Rather good	Normal	Good	Rather good	Normal	Good
	Spread	Rather good	Rather good	Good	Normal	Normal	Good
	Sliminess	Slight	Rather slimy	Slimy	Weak	Rather strong	Strong
	Lubrication	Slightly rough but good	Slightly coarse, united hair	Good	Rather coarse	Coarse	Good
	Softness	Not soft	Very soft	Very soft	Soft	Rather very soft	Very soft
While rinsing	Retained sliminess (after combing 10 times)	3 to 4 times	2 to 3 times	1 to 2 times	1 to 2 times	2 to 3 times	3 to 4 times
	Lubrication	Rather good	Very good	Rather good	Rather good	Slightly rough	Good
	Softness	Moist	Moist	Soft	Slight	None	Flexible
	Hair texture	Coated	Oily	Light and lubricating	Light	Springy	Slightly lubricating
While towel drying	Lubrication	Lubricating but rather rough	Rather rough but good	Rather good	Rough and not good	Rather good	Good
	Softness, hardness	Rather soft	Soft	Intermediate	Rather soft	Flexible	Soft
	Moistness, silkiness	Moist	Moist	Rather moist	Rather moist	Rather moist	Very moist
	Adhesiveness	Adhesive	Rather adhesive	Rather adhesive	Not adhesive	Not adhesive	Slightly adhesive
After drying	Lubrication	Light	Soft	Light	Appropriate	Appropriate	Good
	Softness, hardness	Light	Soft	Soft	Soft	Rather hard	Airy and soft
	Hair texture	Coated	United hair	Smooth	Silky	Firm	United hair

Table 5.13 Sensory characteristics of oils in hair conditioners

Castor oil, vegetable oils and fats, lanolin	Moist, oily
Liquid vegetable oils	Soft
Silicones, straight-branched chain esters	Lubricating
Silicones, straight-branched chain esters, amino modified silicones	Smooth
Glyceryl monostearate, glycols, paste-form oils and fats	Slimy
Beeswax, sterol derivatives, lanolin, dimer esters, high-melting-point waxes	Adhesive
Myristic acid, esters of myristyl alcohol, branched esters of relatively short alkyl group	Silky
Hydrocarbons, high-melting-point solids	Light
High-melting-point waxes, microcrystalline wax	Springy, hard

5.3.5 Prescriptions of Conditioners of Various Sensory Characteristics

Prescriptions of hair conditioners are exemplified below.

5.3.5.1 Coating and Moisturizing Conditioner

Steartrimonium chloride, behetrimonium chloride, and sisteartrimonium chloride are combined to enhance the coated, moist and smooth texture given by dimer dilinoleyl dimer dilinoleate glyceryl tri-hydrogenated rosinate (Prescription 5.16).

Prescription 5.16 Coating and moisturizing conditioner

	Part	Ingredient	%(100 g)
1	A	Steartrimonium chloride (70%)	0.80
2	A	Behetrimonium chloride (80%)	1.60
3	A	Disteartrimonium chloride (75%)	0.90
4	A	Cetanol	4.80
5	A	Glyceryl stearate	1.20
6	A	Microcrystalline wax	2.40
7	A	Ethylhexyl palmitate	0.80
8	A	Dimer dilinoleyl dimer dilionoleate, glyceryl tri-hydrogenated rosinate *3	1.60
9	A	Propyl paraben	0.10
10	A	Methyl paraben	0.10
11	B	Dimethicone 10cs, dimethicone (highly polymerized, 10%)	2.00
12	B	Dimethicone 1000cs	1.00
	C	Purified water	to 100 g

Directions
1) Heat 1–10 to 80°C and dissolve. (A)
2) Heat purified water to 80°C, mix with A and emulsify. (C+A)
3) Cool to 50°C or below, add B and water. (C+A+B)

Ingredient
*3 Lusplan DA-R (Nippon Fine Chemical Co., Ltd.)

5.3.5.2 Moisturizing and Soft Conditioner

Behetrimonium chloride is used as the cationic surfactant, and higher alcohols of oleyl alcohol and behenyl alcohol are added to enhance the moist and soft feeling (Prescription 5.17).

Prescription 5.17 Moisturizing and soft conditioner

		Ingredient	% (100 g)
1	A	Behetrimonium chloride (80%)	2.80
2	A	Cetanol	4.20
3	A	Glyceryl stearate	1.20
4	A	Ethylhexyl palmitate	0.80
5	A	Oleyl alcohol	0.60
6	A	Behenyl alcohol	0.60
7	A	Macadamia nut oil	1.80
8	A	Propyl paraben	0.10
9	A	Methyl paraben	0.10
10	B	Aminoethyl aminopropyl dimethicone copolymer	0.80
11	B	Dimethicone 10cs, dimethicone (highly polymerized, 10%)	2.00
12	B	Dimethicone 1000cs	1.00
	C	Purified water	to 100 g

Directions
1) Heat 1–9 to 80°C and dissolve. (A)
2) Heat purified water to 80°C, mix with A and emulsify. (C+A)
3) Cool to 50°C or below, add B and water. (C+A+B)

5.3.5.3 Lubricating Conditioner

Dicocotrimonium chloride and steartrimonium chloride are combined to enhance lubrication (Prescription 5.18).

Prescription 5.18 Lubricating conditioner

		Ingredient	% (100 g)
1	A	Dicocotrimonium chloride (75%)	0.80
2	A	Steartrimonium chloride (70%)	1.80
3	A	Glyceryl stearate	1.20
4	A	Cetanol	5.60
5	A	Olive oil	1.20
6	A	Octyldodecanol	0.60
7	A	Cetyl ethylhexanoate	1.20
8	A	Propyl paraben	0.10
9	A	Methyl paraben	0.10
10	B	Aminoethyl aminopropyl dimethicone copolymer	0.80

(continued)

Prescription 5.18 (continued)

		Ingredient	% (100 g)
11	B	Dimethicone 10cs, dimethicone (highly polymerized, 10%)	3.00
12	B	Dimethicone 1000cs	1.00
	C	Purified water	to 100 g

Directions
1) Heat 1–9 to 80°C and dissolve. (A)
2) Heat purified water to 80°C, mix with A and emulsify. (C+A)
3) Cool to 50°C or below, add B and water. (C+A+B)

5.3.5.4 Silky Conditioner

Paraffin and octyldodecanol are added to steartrimonium chloride and behetrimonium chloride to give light and silky feeling (Prescription 5.19).

Prescription 5.19 Silky conditioner

		Ingredient	% (100 g)
1	A	Steartrimonium chloride (70%)	1.80
2	A	Behetrimonium chloride (80%)	1.20
3	A	Glyceryl stearate	1.20
4	A	Cetanol	4.80
5	A	Paraffin	1.80
6	A	Octyldodecanol	0.60
7	A	Isopropyl myristate	1.20
8	A	Propyl paraben	0.10
9	A	Methyl paraben	0.10
10	B	Dimethicone 10cs, dimethicone (highly polymerized, 10%)	3.00
11	B	Dimethicone 1000cs	1.00
	C	Purified water	to 100 g

Directions
1) Heat 1–9 to 80°C and dissolve. (A)
2) Heat purified water to 80°C, mix with A and emulsify. (C+A)
3) Cool to 50°C or below, add B and water. (C+A+B)

5.3.5.5 Volume-Building Conditioner

Cetrimonium chloride, steartrimonium chloride, paraffin, dimer dilinoleyl dimer dilinoleate glyceryl tri-hydrogenated rosinate and amino-modified silicone are combined to make the hair firm and still lubricious (Prescription 5.20).

Prescription 5.20 Volume-building conditioner

		Ingredient	%(100 g)
1	B	Cetrimonium chloride (70%)	1.20
2	A	Steartrimonium chloride (70%)	1.80
3	A	Cetanol	4.80
4	A	Ceteth-30	0.40
5	A	Paraffin	2.40
6	A	Dimer dilinoleyl dimer dilionoleate, glyceryl tri-hydrogenated rosinate *3	0.80
7	A	Cetyl ethylhexanoate	1.20
8	A	Propyl paraben	0.10
9	A	Methyl paraben	0.10
10	A	Dimethicone 10cs, dimethicone (highly polymerized, 10%)	3.00
11	B	Dimethicone 1000cs	1.00
	C	Purified water	to 100 g

Directions
1) Heat 1–10 to 80°C and dissolve. (A)
2) Heat purified water to 80°C, mix with A and emulsify. (C+A)
3) Cool to 50°C or below, add B and water. (C+A+B)

Ingredient
*3 Lusplan DA-R (Nippon Fine Chemical Co., Ltd.)

5.3.5.6 Smooth Conditioner

Cetrimonium chloride, steartrimonium chloride, olive oil, dimer dilinoleyl dimer dilinoleate glyceryl tri-hydrogenated rosinate and amino-modified silicone are combined to enhance the smooth and lubricious texture (Prescription 5.21).

Prescription 5.21 Smooth conditioner

		Ingredient	% (100 g)
1	A	Cetrimonium chloride (70%)	1.20
2	A	Behetrimonium chloride (80%)	1.80
3	A	Glyceryl stearate	1.20
4	A	Cetanol	4.80
5	A	Olive oil	1.80
6	A	Dimer dilinoleyl dimer dilionoleate, glyceryl tri-hydrogenated rosinate *3	0.60
8	A	Cetyl ethylhexanoate	1.20
9	A	Lanolin	0.60
10	A	Propyl paraben	0.10
11	B	Methyl paraben	0.10
12	B	Aminoethyl aminopropyl dimethicone copolymer	0.80
13	B	Dimethicone 10cs, dimethicone (highly polymerized, 10%)	3.00
14	B	Dimethicone 1000cs	1.00
	C	Purified water	to 100 g

(continued)

Prescription 5.21 (continued)

Directions
1) Heat 1–10 to 80°C and dissolve. (A)
2) Heat purified water to 80°C, mix with A and emulsify. (C+A)
3) Cool to 50°C or below, add B and water. (C+A+B)
Ingredient
*3 Lusplan DA-R (Nippon Fine Chemical Co., Ltd.)

5.4 Skin Creams

The components of skin creams should be decided based on the aimed sensory characteristics, i.e. for making moisturizing or light cream. The sensory characteristics of creams are determined by the oily components, emulsifiers, polymers, glycols, product form, and viscosity. The sensory characteristics and viscosity can be estimated from the alkyl groups of the oily components and surfactants.

This section describes the effects of the alkyl groups on the feel of use of creams and methods of using polymers, glycols and efficacious and conceptual ingredients.

5.4.1 Constituents of Skin Creams

Skin creams consist of oils, surfactants, fatty acids, alkali agents, moisturizers, and polymers.

Oils to use should be determined first and combined so as to give the moisturizing or light feel to the cream. Oils, such as carbohydrates, esters and higher alcohols, are the main targets of investigation. At least four oils of different melting points and polarities need to be combined. Fatty acids such as stearic acid affect the form and pH of the cream and its efficacious components. Alkali agents as well as moisturizers, which are commonly glycols and saccharides, should also be selected based on the aimed sensory characteristics (either moisturizing or light). Polymers are indispensable for securing the stability of the cream and need to be thoroughly investigated since they affect the properties, external appearance, and viscosity.

5.4.2 Investigation of Oily Constituents

Oils determine the general feel of use and should be investigated first. Carbohydrates, esters, and higher alcohols are the main targets. Vegetable oils and fats should be avoided as much as possible because they are prone to oxidization.

To achieve a light texture, carbohydrates that have a relatively small molecular weight and a branched-chain structure, such as squalane and liquid isoparaffin, should be used. Esters should be liquid esters that have a branched chain in the fatty acid or alcohol group, such as cetyl ethylhexanoate and isopropyl myristate. Higher alcohol should be myristyl alcohol or isostearyl alcohol (Prescription 5.22a).

On the other hand, heavy and moisturizing creams are made by combining liquid paraffin as the carbohydrate component, an ester of relatively large molecular weight, such as cetyl isostearate and trioctanoin, cetanol and stearyl alcohol (Prescription 5.22b).

For producing light and non-slippery creams, solid oils such as paraffin wax, palmityl palmitate, myristyl myristate, and butyl stearate are combined with some liquid oils (Prescription 5.22c).

Fatty acids and higher alcohols have large effects over the viscosity of the creams. Stearyl and cetyl groups increase the viscosity; and thick creams can be produced by increasing the amounts of stearate and cetanol; and the viscosity can be reduced by decreasing cetanol and stearyl alcohol or adding octyldodecanol and/or isostearyl alcohol. Carbohydrates and esters have little effects on viscosity.

To secure creams to be stable at high temperatures, saturated higher alcohols are better combined, such as myristyl alcohol, cetanol, stearyl alcohol, and behenyl alcohol. Steric acid also improves stability.

If vegetable oil is to be combined, oil of low iodine number needs to be selected, and the amount should be kept small. Creams, which are produced by neutralizing fatty acids with alkalis, have neutral to slightly alkaline pH. Vegetable oils are prone to oxidization and smelling at high pHs. The pH should be lowered as much as possible and kept below 5 for combining a large amount of vegetable oils. Vitamin E, lecithin, and other antioxidants may be used but are not very effective in preventing oxidization because they raise the pH.

Prescription 5.22 Skin creams (comparison of sensory characteristics by oil)

			(a)	(b)	(c)
	Part	Ingredient	% (100 g)	% (100 g)	% (100 g)
1	A	Stearic acid	5.60	5.60	5.60
2	A	Beeswax	–	–	1.20
3	A	Glyceryl stearate	1.20	–	1.20
4	A	Cetyl palmitate	–	–	2.40
5	A	Cetyl ethylhexanoate	2.40	4.80	1.60
6	A	Myristyl alcohol	0.60	–	–
7	A	Cetostearyl alcohol	1.80	–	–
8	A	Behenyl alcohol	–	2.40	3.60
9	A	Squalane	0.60	2.40	0.60
10	A	Tocopherol	0.10	0.10	0.10
11	A	Dimethicone 6cs	1.20	1.20	1.20
12	A	PEG-40 stearate	1.20	1.20	1.20
13	A	PEG-25 stearate	0.40	0.40	0.40
14	A	Methyl paraben	0.20	0.20	0.20

(continued)

Prescription 5.22 (continued)

	Part	Ingredient	(a) % (100 g)	(b) % (100 g)	(c) % (100 g)
15	A	Propyl paraben	0.10	0.10	0.10
16	Acidity regulator	10% sodium hydroxide	2.00	2.00	2.00
17	B	Carboxyvinyl polymer	0.20	0.20	0.20
18	B	Xanthan gum	0.10	0.10	0.10
19	B	Glycerin	3.00	4.00	2.00
20	B	BG	3.00	3.00	2.00
		Purified water	76.30	72.30	74.30

Directions
1) Heat 1–15 to 80°C, dissolve, and homogenize. (A)
2) Heat purified water to 80°C. (B)
3) Add B to A, add 16 and emulsify. (B + A)
4) Cool and add water

(a) Light, (b) moist, and (c) nongreasy and non-slippery creams

5.4.3 Selection of Emulsifiers

Nonionic surfactants used in creams include sorbitans, polyglycerins, PEG esters of fatty acids, PEG alkyl ethers, and sucrose esters. A combination of three nonionic surfactants of different HLBs (over 15, around 10 and below 8) makes a good emulsifier. PEG castol and PEG sorbitan oleate can also be used. The combination of surfactants should be determined based on the properties and sensory characteristics of the cream.

Nonionic surfactants that have PEG as the hydrophilic group show drops in detergency at high temperatures since oxygen atoms move away from each other, allowing the carbon atoms to appear on the surface, which makes the chain hydrophobic. Therefore, they lose the emulsifying performance and destabilize the creams. The phenomenon appears as the clouding point and is a disadvantage of PEG surfactants. Although nonionic surfactants are inferior surfactants at high temperatures, they are effective emulsifiers at low temperatures.

Combinations of emulsifiers are summarized below.

5.4.3.1 Fatty Acids Plus Nonionic Surfactants

Fatty acids plus nonionic surfactants are widely used combinations. Most widely used fatty acid is stearic acid, which saponifies and acts as an emulsifier by partial neutralization with alkali. Combined with an alkali agent, such as sodium hydroxide, potassium hydroxide, and triethanol amine, stearic acid makes a slightly alkaline cream. Fatty acid salts undergo reductions in viscosity and may separate and destabilize the cream at high temperatures. To prevent the phenomenon, it is recommended to investigate addition of cetanol or behenyl alcohol as thickeners and carboxyvinyl polymer and xanthan gum as stabilizers.

Widely used nonionic surfactants include sorbitan esters of fatty acids, PEG esters of fatty acids, PEG glycerin esters of fatty acids, and PEG alkyl ethers.

A combination of stearic acid and PEG stearate allows easy emulsification and produces creams of lusterous appearance. It works at a wide HLB range and can be used to emulsify various oils of various polarities without the process of adjusting the HLB to a certain extent. However, it is difficult to produce low-viscosity creams because the alkyl group is stearic acid.

Stearic acid has long been combined with nonionic surfactants of PEG sorbitan and sorbitan esters of fatty acids. Creams of any viscosity and sensory characteristics can be produced from the same oily component by changing the alkyl groups of the surfactants. Surfactants that have oleic acid as the alky group produce low viscous creams, and those that have stearic acid produce thick creams (Prescription 5.23).

Prescription 5.23 Creams (comparison of viscosity by the alkyl group of sorbitan surfactants)

	Part	Ingredient	% (100 g)	% (100 g)	% (100 g)	% (100 g)
1	A	Stearic acid	5.40	5.40	5.40	5.40
2	A	Cetyl ethylhexanoate	4.00	4.00	4.00	4.00
3	A	Cetyl palmitate	0.80	0.80	0.80	0.80
4	A	Squalane	1.80	1.80	1.80	1.80
5	A	Behenyl alcohol	2.40	2.40	2.40	2.40
6	A	Polysorbate 80	2.40	–	2.40	–
7	A	Polysorbate 60	–	2.40	–	2.40
8	A	Sorbitan oleate	0.80	0.80	–	–
9	A	Sorbitan stearate	–	–	0.80	0.50
10	A	Dimethicone 6cs	0.80	0.50	0.50	0.50
11	A	Methyl paraben	0.20	0.10	0.10	0.10
12	A	Propyl paraben	0.10	0.10	0.10	0.10
13	A	Glycerin	2.40	3.00	3.00	3.00
14	Acidity regulator	10% sodium hydroxide	2.00	2.00	2.00	2.00
	B	Purified water	to 100 g	to 100 g	to 100 g	to 100 g
	Viscosity	No. 4 12 rpm, 30 s	7000	8000	10000	13500
	pH		7.20	7.20	7.20	7.20

Directions
1) Heat 1–13 to 80°C, dissolve, and homogenize. (A)
2) Heat purified water to 80°C. (B)
3) Add B to A, add 14 and emulsify. (B + A)
4) Cool and add water

5.4.3.2 Anionic Surfactants Plus Nonionic Surfactants

Combinations of an anionic surfactant plus PEG nonionic surfactants also act as emulsifiers. Anionic surfactants can be sodium cetyl phosphate, sodium myristoyl methyltaurate, sodium cetyl methyltaurate, and sodium isostearoyl lactate. Because lauryl and cocoyl groups are not used, there are few C16 and C18 surfactants left (Prescription 5.24). The nonionic surfactants used combined with fatty acids can also be combined with the anionic surfactants. It should be noted that combinations with anionic surfactants may result in inferior texture and luster of the cream compared to the combinations with stearic acid. This can be solved by adding a small amount of monoglyceryl stearate, sorbitan oleate, or sorbitan stearate. Particularly PEG cetyl ether is prone to producing non-lustrous creams.

Today there are demands for skin creams consisting of only food additives. In such a case, a combination of sodium isostearoyl lactate and sucrose stearate can be used as the emulsifier (Prescription 5.25).

Prescription 5.24 Cream (combination of anionic and nonionic surfactants)

	Part	Ingredient	% (100 g)
1	A	Sodium isostearoyl lactate	4.00
2	A	Cetyl ethylhexanoate	4.00
3	A	Cetanol	6.00
4	A	Squalane	1.00
5	A	PEG-15 glyceryl stearate	2.00
6	A	Glyceryl stearate	1.00
7	A	Dimethicone 6cs	1.00
8	A	Methyl paraben	0.20
9	A	Propyl paraben	0.10
10	Acidity regulator	10% Sodium hydroxide	1.00
11	B	Xanthan gum	2.00
12	B	Glycerin	4.00
13	B	BG	0.20
	B	Purified water	to 100 g
	Viscosity		High
	Sensory		Retained lubrication
	pH		6.4

Directions
1) Heat 1–9 to 80°C, dissolve, and homogenize. (A)
2) Disperse 13 into 11 and 12, add water, heat to 80°C and dissolve. (B)
3) Add B to A, add 10 and emulsify. (B + A)
4) Cool and add water

Prescription 5.25 Creams of food additive surfactants. (Sodium isostearoyl lactate and sucrose ester)

		Ingredient	% (100 g)
1	A	Xanthan gum	0.20
2	A	Sucrose stearate (*4)	1.20
3	A	Glycerin	2.00
4	A	BG	2.00
5	B	Sodium isostearoyl lactate	0.60
6	B	Glyceryl stearate	2.40
7	B	Behenyl alcohol	0.80
8	B	Cetyl ethylhexanoate	3.00
9	B	Sesame oil	1.20
10	B	Tocopherol	0.10
11	B	Dimethicone 6cs	1.20
12	B	Methyl paraben	0.20
13	B	Propyl paraben	0.10
14	pH	50% Lactic acid	
		Purified water	85.00

Directions
1) Disperse 1 and 2 into 3 and 4, add into purified water, heat to 80°C and dissolve. (A)
2) Heat 5–13 to 80°C, and dissolve. (B)
3) Add B to A, add 10 and emulsify. (B + A)
4) Cool to 45°C
5) Adjust the pH with 14, and add water

Specifications
pH: 5.0–6.0

Ingredient
*4: SURFHOPE C-1816 (Mitsubishi Kagaku Foods)

Note: Sucrose stearate is difficult to disperse or dissolve in oil and is prone to form lumps

5.4.3.3 Nonionic Surfactants

To produce creams of low pH for including efficacious and effective ingredients, nonionic surfactants alone should be used as the emulsifier. Anionic surfactants cannot be used for such creams because stearic acid and carboxyvinyl polymer require sodium hydroxide, potassium hydroxide, or triethanol amine as a neutralizer, which increase the pH above 7. PEG nonionic surfactants are vulnerable to destabilization at high temperatures. Xanthan gum plus acrylic acid/(C10–30) alkyl acrylate cross polymer is an effective combination (Prescription 5.26).

Prescription 5.26 Acidic cream (low pH cream of xanthan gum and sorbitan surfactant)

		Ingredient	% (100 g)
1	A	Behenyl alcohol	4.00
2	A	Myristyl alcohol	2.00
3	A	Octyldodecanol	1.00
4	A	Squalane	1.00

(continued)

Prescription 5.26 (continued)

		Ingredient	% (100 g)
5	A	Cetyl ethylhexanoate	5.00
6	A	Polysorbate-60	0.80
7	A	Glyceryl stearate	2.00
8	A	Sorbitan oleate	0.60
9	A	Methyl paraben	0.20
10	A	Propyl paraben	0.10
11	Add afterwards	1% Sodium hyaluronate	0.50
12	B1	Xanthan gum	0.30
13	B1	BG	6.00
14	B1	Glycerin	3.00
15	B2	Citric acid	0.30
16	B2	Sodium citrate	0.10
		Purified water	73.10

Directions
1) Heat 1–10 to 80°C, dissolve, and homogenize. (A)
2) Disperse 12 into 13 and 14. (B1)
2) Dissolve 15 and 16 into purified water, Add B1 and heat to 80°C. (B)
3) Mix A and B, and emulsify. (A+B)
4) Cool, add 12 and water

Specifications
pH: 4.0–5.0

5.4.4 *Addition of Polymers*

Creams consisting of only emulsifiers are prone to viscosity drops, separation and destabilization at high temperatures. Carboxyvinyl polymer, xanthan gum, polyacrylate, acrylic acid/(C10–30) alkyl acrylate cross polymer and other polymers are combined as stabilizers of creams. Polymers undergo little changes in viscosity and thus stabilize creams. Care should be taken to select polymers that do not give heavy or sticky feeling and do not produce gas when applied on the skin (Prescription 5.27).

Prescription 5.27 Cream (nonionic surfactants and polymer stabilizer)

	Part	Ingredient	% (100 g)	% (100 g)
1	A	Carboxyvinyl polymer	0.80	–
2	A	Ammonium acryloyl dimethyl taurate/VP copolymer	–	1.00
3	A	Glycerin	3.00	3.00
4	A	1,3-butylene glycol	3.00	3.00
5	B	Polysorbate-60	1.20	1.20
6	B	Sorbitan oleate	0.40	0.40
7	B	Squalane	2.40	2.40

(continued)

Prescription 5.27 (continued)

	Part	Ingredient	% (100 g)	% (100 g)
8	B	Jojoba oil	0.40	0.40
9	B	Octyldodecanol	0.60	0.60
10	B	Behenyl alcohol	0.80	0.80
11	B	Cetyl ethylhexanoate	1.20	1.20
12	B	Tocopherol	0.10	0.10
13	B	1,2-Pentanediol	0.90	0.90
14	B	Glyceryl caprylate	0.90	0.90
15	Add afterwards	1% Sodium hyaluronate	2.00	2.00
16	Acidity regulator	10% Potassium hydroxide	3.00	–
	A	Purified water	to 100 g	to 100 g

Directions
1) Disperse 1 or 2 into purified water by stirring, add 3 and 4 and heat to 70°C. (A)
2) Heat 5 to 14 to 70°C, dissolve, and homogenize. (B)
3) Add B to A, add C and emulsify. (A+B+C)
4) Add 15 and water.

Specifications
pH: 6.0–8.0

5.4.5 *Selection of Glycols*

Glycols are moisturizing ingredients in creams. Of glycerin, 1,3-butylene glycol and propyrene glycol, glycerin is felt most moisturizing, followed by 1,3-butylene glycol and propyrene glycol in this order. Isoprene glycol gives a more lubricious texture compared to 1,3-butylene glycol. PEG/PPG/polybutylene glycol is a hydrophobic derivative of alkylene oxide and gives a moisturizing effect different from the other glycols and a smooth texture.

5.4.6 *Formulation of Low-Viscosity Creams*

Low-viscous and fluid milky lotions are produced by reducing the amount of stearic acid and using nonionic surfactants to emulsify oils. The oil content should be low; and squalane and octyldodecanol should be added to reduce the viscosity, prevent thickening accompanying aging, and secure stability. Xanthan gum or low-viscous carboxyvinyl polymer is indispensable for securing stability at high temperatures. Polyacrylate, acrylic acid/(C10–30) alkyl acrylate cross polymer, which is a low-viscous emulsification polymer, is effective for reducing the amount of surfactants to make milky lotion of light texture. Suitable alkali agents are potassium hydroxide and triethanol amine, which do not increase viscosity (Prescription 5.28).

Prescription 5.28 Milky lotion (low-viscosity cream)

	Part	Ingredient	% (100 g)	% (100 g)
1	A	Stearic acid	3.00	3.00
2	A	Glyceryl stearate	1.20	1.20
3	A	Cetyl ethylhexanoate	3.60	3.60
4	A	Squalane	1.20	1.20
5	A	Behenyl alcohol	0.80	0.80
6	A	Octyldodecanol	0.60	0.60
7	A	PEG-40 stearate	0.60	0.60
8	A	PEG-25 stearate	0.20	0.20
9	A	Dimethicone 6cs	0.80	0.80
10	A	Methyl paraben	0.10	0.10
11	A	Propyl paraben	0.10	0.10
12	B	Acrylic acid/(C10-30) alkyl acrylate cross polymer	0.20	–
13	B	Xanthan gum	–	0.20
14	B	BG	5.00	5.00
15	Acidity regulator	10% Potassium hydroxide	1.80	1.60
16	B	Purified water	to 100 g	to 100 g

Directions
1) Heat 1–11 to 70°C, dissolve, and homogenize. (A)
2) Disperse 12 (or 13) in 14, add water, heat to 70°C, and dissolve. (B)
3) Add A to B, add 15 and emulsify. (A+B)
4) Cool to 45°C
5) Adjust the pH with 15 and add water

5.4.7 Formulation of Gel Creams

Gel creams can be produced by combining about 1% high-viscous carboxyvinyl polymer and reduced amounts of surfactants and oils as in milky lotions. A small amount of xanthan gum can be included to make smooth and elastic gels. Suitable agents for neutralizing carboxyvinyl polymer are sodium hydroxide, which increases viscosity, and potassium hydroxide and triethanol amine for products of low viscosity. The amounts of oils and surfactants should be reduced compared to ordinary creams. Oils, surfactants, and glycols should be selected so as to achieve the aimed sensory characteristics (Prescription 5.29).

Prescription 5.29 Gel cream (light cream with polymers)

	Part	Ingredient	% (100 g)
1	A	Carboxyvinyl polymer	0.80
2	B	Glycerin	4.00
3	B	Xanthan gum	0.10
4	C	Disodium glycyrrhizate	0.10
5	C	Betaine	0.50
6	C	Trehalose	0.10

(continued)

Prescription 5.29 (continued)

	Part	Ingredient	% (100 g)
7	D	BG	4.00
8	D	1,2-Pentanediol	0.90
9	D	Glyceryl caprylate	0.90
10	D	PEG-60 hydrogenated castol oil	0.60
11	D	Squalane	1.20
12	D	Cetyl ethylhexanoate	0.40
13	D	Tocopherol	0.10
14	D	Sodium dilauramidoglutamide lysine (30%)	0.10
15	E	1% Sodium hyaluronate	2.00
16	E	PCA-Na	0.50
17	F	10% Potassium hydroxide	6.00
	A	Purified water	to 100 g

Directions
1) Disperse 1 into purified water by stirring until uniform, and heat to 70°C. (A)
2) Disperse 2 into 3. (B)
2) Heat 7–14 to 70°C, dissolve, and homogenize. (C)
3) Add B to A, dissolve C, add D and E and homogenize. (A+B+C+D+E)
4) Add F, neutralize, and gelatinize. Add water. (A+B+C+D+E+F)

Specifications
pH: 6.0–8.0

5.4.8 Formulation of Massage Creams

Creams that retain lubrication, smoothness, and spreading performance can be produced by combining a large amount of oils in general, with reduced amounts of solid oils and ingredients of high melting points and increased amounts of liquid oils. Liquid paraffin and natural oils result in retained smoothness on the skin. Inclusion of dimethicones of 50–1000 cs is also effective (Prescription 5.30).

Prescription 5.30 Massage cream (retained lubrication and smoothness)

	Part	Ingredient	% (100 g)
1	A	Stearic acid	4.00
2	A	Monoglyceryl stearate	2.00
3	A	Cetanol	6.00
4	A	Cetyl ethylhexanoate	4.00
5	A	Mineral oil	12.00
6	A	Sorbitan oleate	1.00
7	A	Polysorbate-80	2.00
8	A	Olive oil	4.00
9	A	Dimethicone 50cs	1.00
10	A	Methyl paraben	0.20

(continued)

Prescription 5.30 (continued)

	Part	Ingredient	% (100 g)
11	A	Propyl paraben	0.10
12	C	10% Potassium hydroxide	1.00
13	B	BG	1.00
14	B	Glycerin	6.00
15	B	Xanthan gum	0.20
16	B	Sorbitol (70%)	4.00
	B	Purified water	to 100 g
Viscosity		High	
Sensory		Retained lubrication	
pH		6.4	

Directions
1) Heat 1–11 to 80°C, dissolve, and homogenize. (A)
2) Disperse 15 in 13 and 14, add water, heat to 80°C, dissolve and add 16. (B)
3) Mix A and B, add C and emulsify. (A+B+C)
4) Cool to 45°C
5) Cool and add water

5.5 Leave-On Hair Care Cosmetics

Hair care cosmetics for conditioning and repairing hair from damage contain amino acids, hydrolyzed proteins, ceramides, moisturizing agents, sterols, ingredients for accelerating the penetration and adhesion of the efficacious ingredients to the hair, and components for improving the sensory characteristics. Products for styling and setting hair are designed by combining setting polymers or waxes of high melting points and considering the sensory characteristics and tackiness.

Leave-on hair care cosmetics are applied directly on hair and are not rinsed off, and thus the sensory characteristics of the constituents are directly felt. It is thus important to have thorough knowledge of the sensory characteristics of the ingredients to be combined such as oils, glycerols, and polymers. They are mainly determined by the chemical structure and alkyl and functional groups. Please see Chap. 2.

5.5.1 Hair Creams

Prescription 5.31 is for a hair cream that gives a light texture to the hair. This long-used method emulsifies stearic acid and polyoxyethylene glyceryl monostearate, which is a nonionic surfactant. The appearance and viscosity of the cream can be adjusted by alkali agent for neutralizing stearic acid. The cream of the prescription gives a moisturized feeling. Since it does not contain surfactants, it is difficult to give smooth and soft texture.

Prescription 5.32 is a moisturizing hair liquid. It uses cetyl phosphate and distearic acid, which is a phosphate ester anionic surfactant. Use of a small amount of nonionic surfactant as an emulsification aid helps emulsification and production of lustrous cream. A combination of cetyl phosphate, distearic acid, cholesterol, and hydrolyzed lecithin makes a liquid crystal emulsion. The hair cream is smooth to use unlike those produced by soap emulsification. The pH can be lowered by changing the combination of anionic and nonionic surfactants.

Prescription 5.33 combines anionic surfactant to silicone and nonionic surfactant, and includes small amounts of higher alcohol and branched-chain esters, which are highly compatible to silicone, to help emulsification and improve the stability of the cream. Polyoxyethylene cetyl ether (ceteth) is also added to secure stability. There are various silicone emulsifiers, such as polyoxyethylene and polyoxypropylene. Those that have an alkyl group, such as lauryl, are easy to emulsify. The sensory characteristics of the creams can be changed by using different silicone, esters, and glycols. In this prescription, esters such as dimer dilinoleyl dimer dilinoleate and glyceryl tri-hydrogenated rosinate and a setting polymer of ammonium V/A crotonic acid copolymer are used to obtain hair setting performance as well as silicone for improving the sensory characteristics.

Prescription 5.34 is a hair finishing gel produced by emulsifying silicone with polyoxyethylene cetyl ether and polyoxyethylene lauryl ether sodium acetate (sodium laureth-4 carboxylate). The viscosity and sensory characteristics can be modified by using polyoxyethylene alkyl ether of different alkyl groups.

Silicones are poorly compatible with water and are difficult to produce stable gels. Isodecyl neopentanoate and branched-chain higher alcohols, such as isostearyl alcohol, help emulsification, and improve stability. Addition of esters and higher alcohols, which are highly compatible with water, makes emulsification easy and improves temperature and chronological stability.

The sensory characteristics of oily ingredients are directly reflected on leave-on gels. Dimethicones of low to intermediate viscosity make moisturizing gels, and highly polymerized dimethicones and dimethyl cyclopentasiloxane give light textures. The textures of esters and branched-chain higher alcohols also affect the sensory characteristics as described above.

Prescription 5.35 is a gel cream produced by combining carboxy vinyl polymer, which is anionic acrylate copolymer. The prescription does not use surfactants but uses carboxy vinyl polymer and acrylic acid/(C10-30) alkyl acrylate cross polymer, which is an emulsification polymer and has an alkyl group.

Because no surfactants are used, the cream gives a light feeling. The polymers reduce viscosity changes by temperature and stabilize the cream. Because no cationic surfactants are used, the cream gives a strongly moist texture and is difficult to give a soft feeling. Silicones can be combined to give lubrication.

Oils to be combined can be silicones, esters, etc. and should be determined based on the purpose of use. Glycols should also be selected based on the purpose. Combinations of nonionic and anionic surfactants can also be used to produce gels that give lubrication and luster to the hair by adding a large amount of silicone. Use of vegetable oils and fats should be carefully investigated because they tend to raise the pH of the product.

Prescription 5.31 Hair cream of light texture

		Ingredient	% (100 g)
1	A	PEG-15 glyceryl oleate	0.50
2	A	Stearic acid	4.00
3	A	Glyceryl stearate	1.00
4	A	Myristyl alcohol	2.00
5	A	Phytosteryl isostearate	0.20
6	A	PEG-5 Soy sterol	0.50
7	A	Cetyl ethylhexanoate	4.00
8	A	Squalane	1.00
9	A	Ethylhexyl glycerin	0.20
10	A	Glyceryl caprylate	0.50
11	B	Isopentyldiol	3.00
12	B	10% Potassium hydroxide	1.00
		Purified water	to 100 g
	pH		6.80

Directions
1) Heat 1–11 to 80°C and dissolve. (A)
2) Heat 12 to 80°C and add 11. (B)
3) Mix A and B and emulsify. (A+B)
4) Cool to 45°C and add water

Specifications
pH: 6–7.5

Prescription 5.32 Hair lotion (moisturizing liquid hair cream)

		Ingredient	% (100 g)
1	A	PEG-40 stearate	0.40
2	A	BG	3.00
3	A	Cetostearyl alcohol	1.20
4	A	Glyceryl stearate	0.80
5	A	Dicetyl phosphate, cetyl phosphate, cetostearyl alcohol (*5)	3.20
6	A	Cholesterol	0.20
7	A	Methyl paraben	0.10
8	A	Propyl paraben	0.10
9	A	Cetyl ethylhexanoate	2.40
10	A	Squalane	1.20
11	A	Octyldodecanol	0.40
12	A2	Hydrogenated lecithin	0.20
13	B	10% Potassium hydroxide	1.20
14	C	Silicone emulsion (*2)	5.00
15	pH	50% Lactic acid	
16	B	Purified water	to 100 g

Directions
1) Heat 1–11 to 80°C and dissolve. Add 13 immediately before emulsification. (A)
2) Heat 16 to 80°C and add 13. (B)
3) Mix A and B and emulsify. (A+B)
4) Cool to 45°C, add 14, adjust pH, and add water

Specifications
pH: 4–6

Ingredient
*5 Crodafos CES (Croda Japan)

Prescription 5.33 Hair cream (large silicone content to for easy hair styling)

		Ingredient	% (100 g)
1	A	Ceteth-40	0.80
2	A	Ceteth-2	0.40
3	A	PEG-12 dimethicone	0.60
4	A	Lauryl PEG/PPG-30/10 dimethicone, DPG (*10)	4.00
5	A	Dimer dilinoleyl dimer dilionoleate, glyceryl tri-hydrogenated rosinate (*3)	6.00
6	A	Cyclomethicone dimethiconol	30.00
7	A	Cyclomethicone	30.00
8	B	Triethanol amine	0.90
9	B	Ammonium VA/crotonic acid copolymer	4.00
10	C	Methyl paraben	0.20
11	C	PG	0.80
12	D	Methylisothiazolinone (10%)	0.10
13		Purified water	22.20

Directions
Use homomixer or disperser
1) Weigh 1, 2, 3, and 4 and heat to 50°C. Disperse 5, 6, and 7 uniformly. (A)
2) Add 9–13, and add 8 while stirring. Heat to 70°C and dissolve. (B)
3) Dissolve 10 in 11. (C)
4) Mix A, B, C, and 12 and emulsify. Cool to 40°C. (A+B+C+D)

Ingredients
*3 Lusplan DA-R(Nippon Fine Chemical Co., Ltd.)
*10 BY25-339 Cosmetic Fluid(Dawcorning Toray Co., Ltd.)

Prescription 5.34 Silicone gel (transparent gel)

		Ingredient	% (100 g)
1	A1	Ceteth-2	1.50
2	A1	Ceteth-20	4.50
3	A1	Isostearyl alcohol	1.00
4	A1	Alkyl ethylhexanoate	4.00
5	A1	Isodecyl neopentanoate	4.00
6	A2	Cyclomethicone dimethicone (highly polymerized, 10%)	30.00
7	A2	Dimethicone 50cs	10.00
8	A2	Cyclomethicone	20.00
9	B1	Propylene glycol	0.80
10	B1	Methyl paraben	0.10
11	B1	Propyl paraben	0.10
12	B2	Glycerin	6.00
13	B2	Sodium laureth-4 carboxylate	8.00
14	B2	Sodium benzoate	0.20
15	B2	Purified water	9.80

Directions
Use homomixer or disperser
1) Heat 1, 2, 3, 4, and 5 to 75°C, and homogenize. Add 6, 7, and 8 and heat to 75°C. (A)
2) Add 10 and 11 to 9, heat to dissolve, add 12, 13 14, and 15, and heat to 75°C. (B)
3) Gradually add B into A while stirring A at a high speed. Stir thoroughly. Water is not to be added

Prescription 5.35 Gel cream (lubricating finisher containing emulsification polymers)

		Ingredient	% (100 g)
1	A	Acrylic acid/(C10-30) alkyl acrylate cross polymer	0.30
2	A	Carboxyvinyl polymer	0.40
3	A	Glycerin	3.00
4	B	Propylene glycol	2.00
5	B	Methyl paraben	0.10
6	B	PEG-60 hydrogenated castol oil	0.10
7	B	Phenoxy ethanol	0.50
8	B	Phytosteryl/octyldodecyl *N*-lauroyl-L-glutamate	0.05
9	B	Tocopherol	0.05
10	B	Ethanol	5.00
11	C	PEG-12 dimethicone	0.50
12	C	Dimethicone 1000cs	3.00
13	C	Dimethicone 50 cs	1.00
14	D	Sodium glutamate	0.10
15	D	Arginine	0.10
16	D	EDTA-2Na	0.10
17	D	10% Potassium hydroxide	1.40
18	A	Purified water	to 100 g

Directions
1) Disperse 1 and 2 in 18, add 3 and homogenize. (A)
2) Dissolve 5 and 6 in 4 by heating, add 7, 8, 9, and 10 and homogenize. (B)
3) Mix and homogenize 11, 12, and 13. (C)
4) Dissolve 14, 15, and 16 in 17. (D)
5) Add B to A, add C, homogenize, and add D to make a gel. (A+B+C+D)
6) Adjust pH and add water

Specifications
pH: 5.5–7.0

5.5.2 Cationic Hair Conditioners

Prescription 5.36 is a hair conditioner that uses steartimonium chloride and small amounts of nonionic surfactants, such as PEG stearyl ether as emulsifiers. The conditioner gives a more lubrication and more smooth and favorable texture to the hair than hair creams that contain stearic acid. Steratrimonium chloride is most favored for leave-on hair cosmetics. Cetrimonium chloride gives a light texture, and behetrimonium makes the hair too heavy.

To make a light conditioner, higher alcohols other than cetanol, such as myristyl alcohol, isostearyl alcohol and octyldodecanol, should be combined as well as highly polymerized dimethicone and dimethyl cyclopentasiloxane. Inclusion of dimethicone of intermediate viscosity makes oily and heavy conditioners. Esters also affect the sensory characteristics and thus should be combined as necessary by investigating the structure of the alkyl group. Reducing the amount of higher alcohols for reducing the viscosity may adversely affect the high-temperature stability. Addition of cationized guar gum is effective for preventing the destabilization.

Vegetable oils and fats can be combined by lowering the pH.

To prevent chronological thickening and secure long-term fluidity of the cream, isostearyl alcohol and squalane are combined. Silicone emulsion should be used for combining silicones into low-viscous creams like Prescription 5.36 because oily dimethicone cannot be dispersed.

Prescription 5.37 is a hair conditioner in which steartimonium chloride is combined and emulsified with various cationic surfactants. It is a good prescription for repairing hair damage and protecting hair.

The basis is using cationic surfactants to emulsify the cream. Quaternium-91, isoalkyl (C10–40) amidopropyl ethyl dimmonium ethosulfate, behetrimonium methosulfate, and stearyl dimethylamine as well as steartimonium chloride are good to use.

It is convenient to assume that quaternium-91 is combined for protecting hair and isoalkyl (C10–40) amidopropyl ethyl dimmonium ethosulfate repairs cuticle. Stearyl dimethylamine, which is a tertiary amine, is included as a partner of steartimonium chloride to facilitate adhesion to damaged hair parts, which are anionic.

Highly polymerized dimethicone and amodimethicone give lubrication, and myristyl alcohol and isopropyl myristate give a light texture.

To control spread of hair and help style the hair, dimer dilinoleyl dimer dilionoleate, glyceryl tri-hydrogenated rosinate, VP/dimethylamino ethyl methacrylate copolymer sulfate salt are combined.

Prescription 5.38 is a hair conditioner that contains zein, which is a maize protein. A technique is used to combine zein, which is strongly ionic and does not dissolve in water, into a cationic cream. Zein is easy to dissolve in a mixture of propylene glycol, diproplyene glycol, and water. After it is dissolved in the mixture, trimonium chloride is added to form cationic complexes with zein, which can be uniformly combined in a cationic cream. Also see "Points to note for adding hydrolyzed keratin" in Sect. 5.3.3.9.

Dimer acid esters control spread of hair and facilitate hair styling. This is likely because they are viscous, adhesive, and tacky. The tackiness does not appear directly in the cream but disappears when combined with other oils.

Creams can also be produced by not using cationic surfactants but cationic emulsification polymers. Use of the polymers reduces viscosity changes by temperature (Prescription 5.39). Oily ingredients can be freely combined, and the products are stable and undergo little viscosity changes by temperature. The viscosity can be adjusted by the amount of the cationic polymers. Polymers that have setting performance can also be added without affecting the smooth texture of the cream.

Prescription 5.36 Leave-on liquid hair conditioner (light texture, for bottles)

		Ingredient	% (100 g)
1	A	Steartimonium chloride	1.80
2	A	Cetanol	1.80
3	A	Myristyl alcohol	0.40
4	A	Ethyl hexyl myristate	0.80
5	A	Isostearyl alcohol	0.40
6	A	Squalane	0.20
7	A	Propyl paraben	0.10
8	A	Methyl paraben	0.10
9	A	Steareth-20	0.20
10	C	Silicone emulsion	10.00
11	C	Polyquaternium-64 (*6)	0.10
12	pH	Lactic acid (50% diluted solution)	
13	B	Guar hydroxypropyl trimonium chloride	0.20
14	B	Glycerin	2.00
15	B	PG	3.00
16	B	Purified water	to 100 g

Directions
1) Weigh 1–9 and heat to 80°C and homogenize. (A)
2) Disperse 13 into 14 and 15, and add into 16. Heat to 80°C and dissolve. (B)
3) Mix A and B and emulsify. (A+B)
4) Cool to 45°C. Add 10 and 11. Adjust pH and add water. (A+B+C)

Specifications
pH: 4.0–6.0

Ingredient
*6: Lipidure-C (NOF corporation)

5.5.3 Hair Waxes

Prescription 5.40 is for hair styling waxes that use oily ingredients of high melting points such as microcrystalline wax. The sensory characteristics and styling performance can be adjusted by the melting point of the microcrystalline wax used.

Microcrystalline wax that melts at 88°C gives strong setting performance but makes the hair felt as if entirely coated by resin, adhesive when touched, and poor in lubrication. Microcrystalline wax of melting point of 79°C results in less adhesive touch than the 88°C wax, and gives lubrication and sufficient setting performance. The 70°C microcrystalline wax gives a further reduced setting performance, reduced adhesiveness, and increased smoothness.

When paraffin wax is used instead of microcrystalline wax, the products are moisturizing, smooth and weak in setting performance and are creamy rather than waxy.

Prescription 5.37 Silky hair finisher (creams with cationic surfactants)

		Ingredient	% (100 g)	% (100 g)	% (100 g)	% (100 g)
1	A	Steartimonium chloride (70%)	3.00	3.00	3.00	2.80
2	A	Quaternium-91, cetearyl alcohol, cetrimonium methosulfate (*7)	–	0.50	–	–
3	A	Behetrimonium methosulfate, isoalkyl (C10-40) amidopropyl ethyl dimmonium ethosulfate, cetostearyl alcohol (*8)	–	–	0.50	–
4	A	Stearyl dimethylamine	–	–	–	0.50
5	A	Glyceryl stearate	1.00	1.00	1.00	1.00
6	A	Cetanol	3.00	3.00	3.00	3.00
7	A	Myristyl alcohol	1.00	1.00	1.00	1.00
8	A	Dimer dilinoleyl dimer dilionoleate, glyceryl tri-hydrogenated rosinate	1.00	1.00	1.00	1.00
9	A	Isopropyl myristate	1.00	1.00	1.00	1.00
10	A	Methyl paraben	0.10	0.10	0.10	0.10
11	A	Propyl paraben	0.10	0.10	0.10	0.10
12	A	PG	2.00	2.00	2.00	2.00
13	C	Dimethicone (highly polymerized, 10%), cyclomethicone	8.00	8.00	8.00	8.00
14	C	Amodimethicone, steartrimonium chloride, cocodimonium chloride (*9)	0.60	0.60	0.60	0.60
15	C	Polyquaternium-11 (20%)	2.00	2.00	2.00	2.00
16	pH	50% Lactic acid	0.10	0.10	0.10	0.30
17	B	Purified water	to 100 g	to 100 g	to 100 g	to 100 g

Feel of use				
At application	Moist and light lubrication	Moist and light lubrication	Rather moist and good lubrication	Moist and rather heavy lubrication
At finish	Light and unified hair	Moisturized and unified hair	Moisturized, soft and unified hair	Light, moisturized, smooth and unified hair

Directions

1) Heat 1–12 to 80°C. (A)
2) Heat 17 to 80°C. (B)
3) Mix A and B and emulsify. (A+B)
4) Cool to 45°C, add 13, 14 and 15, adjust the pH with 16, and add water. (A+B+C)

Specifications

pH: 4.0–5.0

Ingredients

*7: CRODAZOSOFT DBQ (CRODA)
*8: CUTISSENTIAL BEHENYL 18-MEA (CRODA)
*9: SM8904 Cosmetic emulsion (Dawcorning Toray Co., Ltd.)

Prescription 5.38 Styling and damage-repair conditioners (zein included, for tubes)

		Ingredient	% (100 g)	% (100 g)	% (100 g)
1	A	Steartrimonium chloride (70%)	2.00	2.00	2.00
2	A	Cetanol	3.00	3.00	3.00
3	A	Glyceryl stearate	0.80	0.80	0.80
4	A	Cetyl ethylhexanoate	1.80	1.80	1.80
5	A	Octyldodecanol	0.20	0.20	0.20
6	A	Propyl paraben	0.10	0.10	0.10
7	A	Methyl paraben	0.10	0.10	0.10
8	A	Dimer dilinoleyl dimer dilionoleate, glyceryl tri-hydrogenated rosinate (*3)	1.80	–	–
9	A	Phytosteryl dimer dilinoleyl dimer dilionoleate	–	1.80	1.80
10	C1	PG	1.00	1.00	1.00
11	C1	Cetrimonium chloride (70%)	0.20	0.20	0.20
12	C2	Water	1.00	1.00	1.00
13	C3	Zein, DPG, water (*2)	1.00	1.00	–
14	C3	Hydrolyzed keratin	–	–	1.00
15	D	Silicone emulsion	5.00	5.00	5.00
16	B	50% Sodium citrate	0.10	0.10	0.10
17	B	Purified water	81.90	81.90	81.90

Directions

1) Weigh 1–12, heat to 80°C and dissolve. (A)

2) Mix 10 and 11, homogenize, add 12, and gradually add 13 and 14. (C)

3) Heat purified water to 80°C, and add 16. Mix with A and emulsify. (B + A)

4) Cool to 45°C, add C, add D and add water. (B + A + C + D)

Specifications

pH: 5.0–6.0

Ingredients

*2: Phytokeluster Z (Ichimaru pharcos Co., Ltd.)

*3: Lusplan DA-R (Nippon Fine Chemical Co., Ltd.)

Prescription 5.39 Smooth hair styling cream (polymers included for emulsification and hair setting)

		Ingredient	% (100 g)	% (100 g)
1	A	Dimethylacrylamide/ethyltrimonium chloride methacrylate copolymer, dipropylene glycol dicaprylate/dicaprate, PPG-1 trideceth-6, (C10, 11) paraffin, sorbitan oleate (*11)	1.60	1.60
2	A	*Olea europaea* (olive) fruit oil	0.50	0.50
3	A	PCA dimethicone	1.00	1.00
4	A	Cyclomethicone	6.00	6.00
5	C	Pentylene glycol	0.80	0.80
6	C	Phenoxy ethanol	0.50	0.50
7	C	PG	2.00	2.00
8	C	PEG polybutylene glycol/PPG-8/5 glycerin	1.00	1.00

(continued)

Prescription 5.39 (continued)

		Ingredient	% (100 g)	% (100 g)
9	D	Methacroyl oxyethyl carboxy betaine acrylate copolymer (*12)	5.00	–
10	D	Polyquaternium-11 (*13)	–	5.00
11	pH	Sodium lactate	0.20	–
12	B	Raffinose	0.20	0.20
13	B	Betaine	0.20	0.20
14	B	Purified water	81.00	81.20
Feel of use			Moisturized and unified hair	Manageable hair ends and hair of firm texture

Directions
1) Homogenize 1–4. (A)
2) Add 12 and 13 to 14, heat to 50°C, and dissolve. (B)
3) Homogenize 5–8. (C)
4) Add A to B and homogenize. Add C, 9 and 10 and homogenize. (B + A + C + D)
5) Adjust pH and add water

Specifications

pH: 3.8–4.6

Ingredient
*11: TINOVIS® CD (Ciba)
*12: RAM Resine 3000 (Osaka Organic Chemical Industry Ltd..)
*13: Gafquat 755 N (ISP)

Prescription 5.40 Hair wax (melting point of microcrystalline wax and sensory characteristics)

		Ingredient	% (100 g)	% (100 g)	% (100 g)	% (100 g)
1	A	Microcrystalline wax (88°C)	3.50	–	–	–
2	A	Microcrystalline wax (79°C)	–	3.50	–	–
3	A	Microcrystalline wax (70°C)	–	–	3.50	–
4	A	Paraffin wax (69°C)	–	–	–	3.50
5	A	Candelilla wax	3.50	3.50	3.50	3.50
6	A	Hydroxystearate	6.00	6.00	6.00	6.00
7	A	Ceteth-2	6.00	6.00	6.00	6.00
8	A	Ceteth-7	2.00	2.00	2.00	2.00
9	A	Ceteth-30	1.00	1.00	1.00	1.00
10	A	Cetyl ethylhexanoate	8.00	8.00	8.00	8.00
11	A	Cetanol	4.00	4.00	4.00	4.00
12	A	Propyl paraben	0.10	0.10	0.10	0.10
13	A	Methyl paraben	0.30	0.30	0.30	0.30
14	A	1,3-Butylene glycol	2.00	2.00	2.00	2.00
15	A	Dimethicone 100cs	1.00	1.00	1.00	1.00
16	B	Sodium benzoate	0.20	0.20	0.20	0.20
17	B	10% Sodium hydroxide	2.40	2.40	2.00	2.00

(continued)

Prescription 5.40 (continued)

		Ingredient	% (100 g)	% (100 g)	% (100 g)	% (100 g)
18	B	Sodium lactate	1.20	1.20	1.20	1.20
19	B	Purified water	58.80	58.80	59.20	59.20

Directions
1) Heat 1–15 to 85°C and dissolve. (It becomes cloudy.) (A)
2) Heat 16, 17 and 18 to 80°C and dissolve. (B)
3) Mix A and B and add water. (A + B)

Specifications
pH: 3.8–4.6

5.5.3.1 Hardness of Hair Waxes and pH Regulation

The hardness of hair waxes, which may be either solid or creamy, is determined by whether hydroxystearate or stearic acid is used (Prescription 5.41). Hydroxystearate gelatinizes liquid oils and/or combined with waxes of high melting points and thus produces hard solid waxes.

The hardness of wax products is also affected by production method. When the emulsion is poured into a container while it is still hot (about 70°C), the cooled product becomes a hard wax. On the other hand, when the emulsion is cooled to about 45°C before pouring into a container, the product becomes a soft cream. Hard waxes can also be produced by reducing the amount of sodium hydroxide and lowering the pH.

Adjusting the pH by only using sodium hydroxide is rather difficult because the buffer power is weak, and thus the pH may sharply change by an addition of a small amount of sodium hydroxide. Combining sodium lactate facilitates pH regulation.

Prescription 5.41 Hair wax (difference in sensory characteristic between stearic acid and hydroxystearate)

		Ingredient	% (100 g)	% (100 g)
1	A	Microcrystalline wax (88°C)	3.50	3.50
2	A	Candelilla wax	3.50	3.50
3	A	Hydroxystaerate	6.00	–
4	A	Stearic acid	–	6.00
5	A	Ceteth-2	6.00	6.00
6	A	Ceteth-7	2.00	2.00
7	A	Ceteth-30	1.00	1.00
8	A	Cetyl ethylhexanoate	8.00	8.00
9	A	Cetanol	4.00	4.00
10	A	Propyl paraben	0.10	0.10
11	A	Methyl paraben	0.30	0.30
12	A	BG	2.00	2.00
13	A	Dimethicone 100 cs	1.00	1.00
14	B	Sodium benzoate	0.20	0.20

(continued)

Prescription 5.41 (continued)

		Ingredient	% (100 g)	% (100 g)
15	B	10% Sodium hydroxide	2.40	2.40
16	B	Sodium lactate	1.20	1.20
17	B	Purified water	58.80	58.80
		pH	7.00	7.00
		Characteristic	Waxy	Creamy

Directions
1) Heat 1–13 to 85°C and dissolve. (It becomes cloudy.) (A)
2) Add 14 to 16 to purified water and heat to 80°C and dissolve. (B)
3) Mix A and B and add water

Specifications
pH: 5.0–7.5

5.5.3.2 High-Temperature Stability of Waxes Containing Hydroxystearate

Hair waxes that contain hydroxystearate are prone to destabilization at high temperatures. They may separate when combined with inappropriate combinations of nonionic surfactants or surfactants of inappropriate HLB. Unlike stearic acid, hydroxystearate does not act as an emulsifier even at the soap state with alkalis. It serves as a gelatinization agent of oils and is thus at least partially dissolved in the oils.

This disadvantage can be solved by adding high molar PEG nonionic surfactants. Nonionic surfactants of PEG larger than 30 improves stability.

5.5.3.3 Hard Wax Not Containing Setting Polymers

Prescription 5.42 is a hair wax of very high hair setting performance, in which high-melting point waxes, setting polymers and esters are combined. Hair styling performance can be achieved even by omitting setting polymers by making the use of the characteristics of the waxes. Addition of setting polymers enhances firmness but increases tackiness.

Prescription 5.42 Wax for swept up hair style

		Ingredient	% (100 g)	% (100 g)	% (100 g)	% (100 g)
1	A	Behenyl alcohol	2.00	2.00	2.00	2.00
2	A	Cetanol	3.00	3.00	3.00	3.00
3	A	Hydroxystearate	4.00	4.00	4.00	4.00
4	A	Glyceryl stearate	2.00	2.00	2.00	2.00
5	A	Microcrystalline wax 155 F	3.00	3.00	3.00	3.00
6	A	Rice bran wax	2.00	2.00	2.00	2.00
7	A	Dimer dilinoleyl dimer dilionoleate, glyceryl tri-hydrogenated rosinate	2.00	2.00	2.00	2.00

(continued)

Prescription 5.42 (continued)

		Ingredient	% (100 g)	% (100 g)	% (100 g)	% (100 g)
8	A	Pentaerythrityl tetra (2-ethylhexanoate)	6.00	6.00	6.00	6.00
9	A	Sodium stearoyl lactate	0.80	0.80	0.80	0.80
10	A	Ethylhexyl glycerin	0.50	0.50	0.50	0.50
11	A	glyceryl caprylate	0.50	0.50	0.50	0.50
12	A2	Acrylates/octryl acrylamide copolymer (*12)	–	4.00	–	–
13	A3	PG	4.00	4.00	4.00	4.00
14	B	Sucrose stearate (*4)	2.00	2.00	2.00	2.00
15	D	TEA acrylates copolymer (*14)	–	–	10.00	–
16	D	AMP-acrylates/diacetoneacrylamide copolymer (*15)	–	–	–	10.00
17	D	Dimethicone (highly polymerized 10%), cyclomethicone	4.00	4.00	4.00	4.00
18	D	Dimethicone 10cs	1.00	1.00	1.00	1.00
19	C	10% Sodium hydroxide	0.60	5.00	0.60	0.60
20	B	Purified water	52.60	74.70	74.70	74.70

Sensory characteristics

Viscosity	Cream	Waxy	Cream	Cream
Feel at application	Easy to spread and apply	Easy to spread and apply	Rather greasy. Spreads easily	Easy to spread and apply
Hair setting performance	Weak	Not lasting	Good for giving luster. Rather greasy	Sufficient setting performance for sweeping up the hair
Tackiness	Tacky	Tacky	Tacky	Tacky

Directions

1) Heat 1–11 to 80°C and dissolve. Disperse 12, add 13, and dissolve. (A)

2) Add 14 to purified water, heat to 80°C and dissolve. (B)

3) Ad A to B, and add 19 and emulsify. (A+B+C)

4) Cool to 45°C. Add 15 to 18, and add water. (A+B+C+D)

Specifications

pH: 6.2–7.4

Ingredients

*4: SURFHOPE C-1816 (Mitsubishi Kagaku Foods)

*14: Amphomer V-42 (AkzoNobel)

*15: Aniset KB-100 H (Osaka Organic Chemical Industry Ltd..)

*16: Plas cize L-53P (Goo Chemical Co., Ltd..)

5.5.3.4 Adjusting the Sensory Characteristics by Esters

The sensory characteristics change by esters (Prescription 5.43). Cetyl ethylhexanoate gives stronger lubrication than C14–18 alkyl ethylhexanoate. Diethylhexyl succinate gives the strongest hair setting performance. Caprylyl eicoseonate also gives good hair setting performance. Adding solid esters of larger molecular weights to diethylhexyl succinate produces harder and weaker lubricating feel.

All esters give adhesive and heavy texture, but some give lighter feel than the others.

5.5.3.5 Low-Viscous Hair Waxes for Women

Most hair waxes are solids or hard creams. Female consumers prefer low-viscous waxes because hard waxes stick to fingers and remain under the nails. Prescription 5.44 is a liquid hair wax produced by emulsifying with acrylic acid/(C10–30) alkyl acrylate cross polymer and polyoxyethylene cetyl ether. Hair setting polymers help hair styling.

5.5.4 Hair Gels

Most hair gels uses carboxy vinyl polymer as the main thickener. Other polymers may be used but they must not kink by rubbing. Gels produced by thickeners become fluid and thixotropic.

There are hair gels for hair styling and setting, which contain hair setting polymers, and those for hair conditioning, which contain oils, hydrolyzed proteins and amino acids. Various hair setting polymers are available including nonionic, anionic, cationic and amphoteric polymers and those of high to weak hair setting performances. Polymers suitable for the purpose should be selected. The sensory characteristics given by major hair setting polymers are summarized in Tables 5.14 and 5.15.

Carboxy vinyl polymer is anionic, but cationic polymers may be combined in some cases. There are also cases in which even amphoteric polymers cannot be combined. Prescription 5.45 is for hair gels of strong hair setting performance and uses octylacrylamide/hydroxypropyl acrylate/butylaminoethyl methacrylate copolymer and two nonionic hair setting polymers (VP/VA copolymer and PVP). Prescription 5.46 is for cationic transparent hair setting finishers.

Prescription 5.43 Hair wax (differences in sensory characteristics by ester)

		Ingredient	% (100 g)	% (100 g)	% (100 g)	% (100 g)	% (100 g)	% (100 g)	% (100 g)
1	A	Ceteth-2	6.00	6.00	6.00	6.00	6.00	6.00	6.00
2	A	Ceteth-7	2.00	2.00	2.00	2.00	2.00	2.00	2.00
3	A	Ceteth-30	1.00	1.00	1.00	1.00	1.00	1.00	1.00
4	A	Stearic acid	5.00	5.00	5.00	5.00	5.00	5.00	5.00
5	A	Microcrystalline wax (70°C)	3.00	3.00	3.00	3.00	3.00	3.00	3.00
6	A	Cadelilla wax	3.00	3.00	3.00	3.00	3.00	3.00	3.00
7	A	Stearyl alcohol	3.00	3.00	3.00	3.00	3.00	3.00	3.00
8	A	Cetyl ethylhexanoate	10.00	–	–	–	–	–	–
9	A	Myristyl ethylhexanoate	–	10.00	–	–	–	–	–
10	A	Diethylhexyl succinate	–	–	10.00	–	6.00	6.00	6.00
11	A	Caprylyl eicosenoate	–	–	–	10.00	–	–	–
12	A	Myristyl myristate	–	–	–	–	4.00	–	–
13	A	Cetyl palmitate	–	–	–	–	–	4.00	–
14	A	Stearyl stearate	–	–	–	–	–	–	4.00
15	A	Propyl paraben	0.10	0.10	0.10	0.10	0.10	0.10	0.10
16	A	Methyl paraben	0.20	0.20	0.20	0.20	0.20	0.20	0.20
17	A	BG	3.00	3.00	3.00	3.00	3.00	3.00	3.00
18	A	Dimethicone 100 cs	1.00	1.00	1.00	1.00	4.00	4.00	4.00
19		Lactic acid (50%) for acidity regulation	1.00	1.00	1.00	1.00	1.00	1.00	1.00
20	B	Sodium hydroxide	0.50	0.50	0.50	0.50	0.50	0.50	0.50
21	B	Purified water	to 100 g	to 100 g	to 100 g	to 100 g	to 100 g	to 100 g	to 100 g

Sensory characteristics							
At application	Light and smooth on the palm	Light and smooth on the palm	Viscous and heavy on the palm	Heavy, moist and viscous on the palm	Viscous, rather hard, and rather slippery on the palm	Viscous, hard and rather slippery on the palm	Viscous, hard and rather slippery on the palm
Hair setting performance	Weakest	Weak	Good	Normal	Rather weak	Weak	Weak
Tackiness	Not tacky	Not tacky	Tacky	Rather tacky	Slightly tacky	Slightly tacky	Slightly tacky
Texture	Most lubricating, light	Light and lubricating, silky	Moist and heavy	Weak lubrication, heavy and moist	Weak lubrication, rather hard	Weak lubrication, hard	Slightly lubricating, hard

Directions
1) Weigh 1–17, heat to about 80°C, and dissolve. (A)
2) Heat purified water to about 80°C and dissolve 20. (B)
3) Mix A and B, and emulsify. Cool to 45°C, and add 18. (A+B+C)
4) Cool to the room temperature. Adjust pH with 19. Add water

Prescription 5.44 Hair wax lotion (liquid hair wax)

		Ingredient	% (100 g)	% (100 g)	% (100 g)	% (100 g)
1	A	Candelilla wax	1.50	1.50	1.50	1.50
2	A	Microcrystalline wax 155 F	0.50	0.50	0.50	0.50
3	A	Cetyl ethylhexanoate	6.00	6.00	6.00	6.00
4	A	*Simmondsia chinensis* (jojoba) seed oil	0.10	0.10	0.10	0.10
5	A	Ceteth-25	1.20	1.20	1.20	1.20
6	A	Ceteth-7	0.40	0.40	0.40	0.40
7	A	Ceteth-8	0.80	0.80	0.80	0.80
8	A	Propyl paraben	0.10	0.10	0.10	0.10
9	A	Dimethicone (highly polymerized, 10%), cyclomethicone	3.00	3.00	3.00	3.00
10	C	Methyl paraben	0.20	0.20	0.20	0.20
11	C	Phenoxy ethanol	0.60	0.60	0.60	0.60
12	C	BG	3.00	3.00	3.00	3.00
13	C	Ethanol	3.00	3.00	3.00	3.00
14	C	VA/crotonates/vinyl neodecanoate copolymer (*17)	–	1.00	–	–
15	D	AMP-polyurethane-14,acrylates copolymer (*18)	–	–	4.00	–
16	D	TEA acrylates copolymer (*15)	–	–	–	3.00
17	E	10% Potassium hydroxide	0.80	0.80	0.80	0.80
18	B	Acrylates/C10-30 alkyl acrylate cross polymer	0.30	0.30	0.30	0.30
19	B	Purified water	to 100 g	to 100 g	to 100 g	to 100 g

Sensory characteristics

Viscosity	Rather pudding-like, low	Pudding-like, high	Low viscosity	Milky
Feel at application	Spreads very well and does not remain on the hand	Spreads well, and gives remaining feeling on the hand	Spreads very well and does not remain on the hand	Spreads rather poorly and gives remaining feeling on the hand
Hair setting performance	Almost none	Normal	Good	Good
Luster	Rather lustrous	Slightly lustrous	Rather lustrous	Almost none
Tackiness	Tacky	Tacky	Tacky	Rather tacky
Texture	Smooth and light	Soft	Rather light	Rather heavy

Directions

1) Heat 1–8 to°C and dissolve. Add 9 immediately before emulsification. (A)
2) Disperse 18 in purified water, heat to 80°C, and homogenize. (B)
3) Dissolve 10 into 11 to 13, and add 14. (C)
4) Mix A and B, add 17, and emulsify
5) Cool to 45°C. Add C, 15 and 16. (A+B+C+D)
6) Adjust pH and add water

(continued)

Prescription 5.44 (continued)

Specifications pH: 6.0–7.5
Ingredients *17: Resyn 28–2930 (AkzoNobel) *18: DynamX (AkzoNobel) *15: Aniset KB-100 H (Osaka Organic Chemical Industry Ltd..)

Table 5.14 Comparison of carbomer gels containing different setting polymers

A Basic prescription

	Ingredient	
	Hair setting polymer	
1	Citric acid	0.20
2	Carbomer	0.60
3	Purified water	70.00
4	EDTA-2Na	0.10
5	Propylene glycol	4.00
6	Methyl paraben	0.20
7	Propyl paraben	0.05
8	PEG-10 methyl ether dimethicone	0.80
9	Ethanol	6.00
10	10% Potassium hydroxide	4.00
11	10% Potassium hydroxide or 10% citric acid	
	Purified water	to 100 g

Directions
1) Dissolve 1 into 3, gradually add 2 to disperse uniformly, and dissolve 4. (A)
2) Add 6 and 7 to 5, heat to 60°C to dissolve, and add 8. (B)
3) Dissolve or dilute the setting polymer with 9, add 10, and stir until uniform. (C)
4) Add B to A, add C, and stir. Adjust pH with 18, and add water. (A+B+C)

Specifications
pH: 6.0–7.5

Table 5.15 Differences in property and sensory characteristic by hair setting polymer

Hair setting polymer	Amount % (100 g)	pH	Appearance	Sensory Spread	Setting	Hair unity	Tackiness
PVP/VA(*1)	4.00	6.9	Transparent lotion	Good	Bad	Bad	Tacky
PVP(*2)	2.00	6.2	Transparent gel	Good	Rather good	Good	Not tacky
Polyquaternium-11(*3)	10.00	7.2	Transparent gel	Good	Strong	Rather good	Rather tacky
Ethyl ester of PVM/MA copolymer(*4)	4.00	6.9	Transparent lotion	Good	Rather tacky	Rather good	Tacky

(continued)

Table 5.15 (continued)

Hair setting polymer	Amount % (100 g)	pH	Appearance	Sensory Spread	Setting	Hair unity	Tackiness
N-Methacryloyl oxyethel *N,N'*-dimethyl ammonium-α-*N*-methyl carboxy betaine and Alkyl methacrylate copolymer(*5)	6.70	7.4	Aggregates				
Acrylates/octylacrylamide copolymer(*6)	2.00	6.5	Milky gel	Good	Good	Good	Slightly tacky
Octylacrylamide/acrylates/butylaminoethyl methacrylate copolymer(*7)	2.00	6.3	Milky gel	Good	Good	Good	Slightly tacky
VA/crotonates/vinyl neodecanoate copolymer(*8)	2.00	6.3	Transparent lotion	Good	Weak	Bad	Rather tacky
Polyurethane-14 AMP-acrylates copolymer(*9)	6.80	6.3	Milky gel	Good	Rather good	Slightly tacky	Tacky
AMP- acrylates copolymer(*10)	5.00	6.5	Milky gel	Good	Good	Good	Slightly tacky
Polyquaternium-49(*11)	4.00	7.0	Transparent gel	Good	Good	Good	Not tacky
Polymethacryloyl ethyl betaine(*12)	6.70	6.4	Aggregates				
AMP- acrylates copolymer(*13)	5.00	6.3	Milky gel	Good	Good	Rather good	Slightly tacky

*1	PVP/VA E-735(50%)	(ISP)	PVP/VA
*2	PVP K30(100%)	(ISP)	PVP
*3	Gafquat 755 N(20%)	(ISP)	Polyquaternium-11
*4	Gantrez SP-215(50%)	(ISP)	Ethyl ester of PVM/MA copolymer
*5	Yukaformer R205S(30%)	(Mitsubishi Kagaku)	*N*-Methacryloyl oxyethel *N,N'*-dimethyl ammonium-α-*N*-methyl carboxy betaine and Alkyl methacrylate copolymer
*6	AMPHOMER V-42(100%)	(Akzonobel)	Acrylates/octylacrylamide copolymer
*7	AMPHOMER 28-4910(100%)	(Akzonobel)	Octylacrylamide/acrylates/butylaminoethyl methacrylate copolymer
*8	Resyn 28-2930(100%)	(Akzonobel)	VA/crotonates/vinyl neodecanoate copolymer
*9	Dynam X(28%)	(Akzonobel)	Polyurethane-14 and AMP-Acrylates Copolymer
*10	Aniset KB-100 H(40%)	(Osaka Organic Chemical Industry Ltd..)	AMP- acrylates copolymer
*11	Wet resin(50%)	(Osaka Organic Chemical Industry Ltd..)	Polyquaternium-49

(continued)

Table 5.15 (continued)

*12	Plascize L-401(30%)	(Goo Chemical Co., Ltd..)	Polymethacryloyl ethyl betaine
*13	Plascize L-8011(40%)	(Goo Chemical Co., Ltd..)	AMP- acrylates copolymer
*14	RAM Resin 3000(30%)	(Osaka Organic Chemical Industry Ltd..)	*N*-Methacryloyl oxyethel *N,N'*-dimethyl ammonium-α-*N*-methyl carboxy betaine and Alkyl methacrylate copolymer
*15	Plas cize L-53P	(Goo Chemical Co., Ltd..)	AMP-acrylates/diacetoneacrylamide copolymer
*16	AMPHOMER SH30(30%)	(Akzonobel)	Octylacrylamide/hydroxypropyl acrylate/butylaminoethyl methacrylate copolymer
*17	Luviquat FC550(40%)	(BASF)	Polyquaternium-16
*18	Styleze W-10(10%)	(ISP)	Polyquaternium-55

Prescription 5.45 Transparent hair setting gel

		Ingredient	% (100 g)	% (100 g)	% (100 g)
1	A	Diethoxyethyl succinate	2.00	2.00	2.00
2	A	1,3-Butylene glycol	2.00	2.00	2.00
3	A	Ethyl alcohol	7.00	7.00	7.00
4	A	Methyl paraben	0.10	0.10	0.10
5	A	Propyl paraben	0.05	0.05	0.05
6	A	Octylacrylamide/hydroxypropyl acrylate/butylaminoethyl methacrylate copolymer (*19)	4.00	4.00	4.00
7	A	VP/VA copolymer (*20)	–	2.40	–
8	A	PVP	–	–	1.20
9	B	Purified water	to 100 g	to 100 g	to 100 g
10	C	TEA	2.00	2.00	2.00
11	D	Acrylates/steaereth-20 itaconate copolymer	7.00	7.00	7.00
12	D	Purified water	16.00	16.00	16.00

Sensory characteristics

Hair setting performance	Weak	Strong	Good
Texture	Rather tacky but not bothering	Very tacky until it dries. Becomes crispy when it dries	Not tacky to rather tacky. Does not become crispy when it dries but gives a unique texture

Directions

1) Homogenize 1–8. (A)

2) Dilute 11 with 12. (B)

3) Add A to water, and add B. Add C gradually while stirring to make a transparent gel

Specifications

pH: 7–8

Ingredients

*19: Amphomer SH30 (AkzoNobel)

*20: PVP/VA E-735 (ISP)

Prescription 5.46 Hair essence (cationic transparent hair setting finisher)

		Ingredient	% (100 g)	% (100 g)	% (100 g)	% (100 g)
1	A	1,3-Butylene glycol	3.00	3.00	3.00	3.00
2	A	PEG/PPG/Polybutylene glycol −8/5/3 glycerin	2.00	2.00	2.00	2.00
3	A	Pentylene glycol	1.00	1.00	1.00	1.00
4	A	Steartrimonium chloride (70%)	0.50	0.50	0.50	0.50
5	A	PEG-60 hydrogenated castol oil	0.20	0.20	0.20	0.20
6	a	Scenting agent	0.10	0.10	0.10	0.10
7	C	Octylacrylamide/ hydroxypropyl acrylate/butylaminoethyl methacrylate copolymer (*22)	–	5.00	–	–
8	C	Polyquaternium-11 (*13)	–	–	5.00	–
9	C	VP/VA copolymer (*20)	–	–	–	5.00
10	C	Sodium coco PG-dimonium chloride phosphate (*21)	0.50	0.50	0.50	0.50
11	C	Ethanol	4.00	4.00	4.00	4.00
12	B	Citric acid	0.20	0.20	0.20	0.20
13	B	Arginine	0.10	0.10	0.10	0.10
14	B	Betaine	0.30	0.30	0.30	0.30
15	B	Creatine	0.10	0.10	0.10	0.10
16	B	Polyquaternium-67	0.30	0.30	0.30	0.30
17	B	Purified water	87.70	82.70	82.70	82.70
pH			5.6	5.8	5.7	5.4
External appearance			Transparent lotion	Transparent lotion	Transparent lotion	Transparent lotion
Sensory characteristics						
At application			Slimy	Slimy and rather lubricating	Slimy and rather lubricating	Slimy and rather lubricating
After drying			Soft and moist. Rather oily and lubricating	Rather bad lubrication. Tacky and oily	Little tacky and good lubrication	Bad lubrication and tacky

(continued)

Prescription 5.46 (continued)

Directions
1) Heat 1–5 to 60°C, and add. (A)
2) Mix 12–16 in form of powder, add purified water, heat to 50°C, and dissolve to make a viscous liquid. (B)
3) Add A to B. Add 7 to 11 separately, and add water. (B + A + C)
Specifications
pH: 5–6
Ingredients
*13: Gafquat 755 N (ISP)
*20: PVP/VA E-735 (ISP)
*21: Arlasilk Phospholipid CDM (Croda Japan)

5.5.4.1 Inclusion of Ceramides, Amino Acids, and Hydrolyzed Proteins

Oils are widely used in hair gels for hair conditioning. Oils should be selected based on the purpose. Efficacious ingredients such as ceramides, amino acids, and hydrolyzed proteins are also widely combined.

Prescription 5.47 is dry finishers that make the hair lustrous. The combination of dimer dilinoleyl dimer dilionoleate, glyceryl tri-hydrogenated rosinate, and highly polymerized dimethicone gives good luster to the hair. Dimer dilinoleyl dimer dilionoleate and glyceryl tri-hydrogenated rosinate are esters of good hair setting performance. The combination of bis-ethoxydiglycol cyclohexane 1,4-dicarboxylate and PEG-60 hydrogenated castor oil PCA isostearate can solubilize highly polar oils with a relatively small amount of surfactants.

Prescription 5.48 is a transparent hair conditioning gel. Ammonium acryloyldimethyl taurate/VP copolymer produces a soft gel and gives almost no feel, allowing the sensory characteristics of the other ingredients to appear.

Prescription 5.49 is a viscous transparent hair gel, which uses the gelatinization characteristics of the nonionic polyoxyethylene ether of relatively high HLB with water.

PEG-40 hydrogenated castol oil and ceteth-40 are used to make the transparent gel. Combining nonionic surfactants, propylene glycol, and water at appropriate ratio makes a gel. Addition of cationic surfactants gives smoothness. The combination of the nonionic surfactant, ester and glycol gives strong luster, a rather heavy but moist touch, and unity to the hair.

Prescription 5.47 Hair gel finisher (for dry and lustrous hair)

		Ingredient	% (100 g)	% (100 g)
1	A	PEG-60 hydrogenated castor oil PCA isostearate	0.90	0.90
2	A	PG	3.00	3.00
3	A	Dipropylene glycol	2.00	2.00
4	A	Phytosteryl macadamiate	0.05	0.20
5	A	Tocopherol	0.05	–
6	A	Bis-ethoxydiglycol cyclohexane 1, 4-dicarboxylate (*22)	1.00	–
7	A	Octyldodecanol	–	0.50
8	A	Dimer dilinoleyl dimer dilionoleate, glyceryl tri-hydrogenated rosinate (*3)	–	2.00
9	A	Glyceryl stearate	–	0.50
10	A	Benzyl alcohol	0.20	0.20
11	B	Carbomer	0.95	0.95
12	C	Betaine	0.50	0.50
13	D	Ethanol	18.00	18.00
14	D	Polyquaternium-61 (*23)	0.20	0.20
15	D	Hydroxypropyl trimonium hyaluronate	3.00	3.00
16	D	PCA-Na	0.50	0.50
17	D	Sodium lactate	0.10	0.10
18	D	Hydrogenated polyisobutene, dimethicone (highly polymerized) (*24)	–	12.00
19	pH	10% Sodium hydroxide	3.80	3.80
	B	Purified water	65.75	51.65
		Characteristics	Transparent gel	Milky gel.
		Sensory characteristics		
		Spread at application	Light and spreads well	Rather moist and spreads well
		Luster	Gives good luster	Gives good luster
		Hair unity	Good unity of hair ends	Good unity of hair ends
		Texture	Soft and silky	Moist and lubricating

Directions
1) Heat 1–10 to 60°C and dissolve. (A)
2) Disperse 11 in purified water. (B)
3) Add 12 to B, and add A. (B+C+A)
4) Add 13–18, adjust pH with 19, and add water. (B+C+A+D)

Specifications
pH: 6–7

Ingredients
*22: Neosolue-Aqulio (Nippon Fine Chemical Co., Ltd.)
*3: Lusplan DA-R (Nippon Fine Chemical Co., Ltd.)
*23: Lipidure-NA (NOF corporation)
*24: BY25-320 (Dawcorning Toray Co., Ltd.)

Prescription 5.48 Transparent hair conditioning gel

		Ingredient	% (100 g)
1	A	Methyl paraben	0.20
2	A	Propyl paraben	0.05
3	A	Isoprentyl glycol	3.00
4	A	PEG-4	2.00
5	A	Jojoba wax PEG-80	0.50
6	A	Ammonium acryloyldimethyl taurate/VP copolymer	1.20
7	C	Sodium dilauramidoglutamide lysine (30%)	0.20
8	C	PCA-Na (50%)	3.00
9	D	Polyquaternium-64 (*6)	0.30
10	D	Purified water	5.00
11	B	Aspartic acid	0.30
12	B	Glutamic acid	0.30
13	B	EDAT-4Na	0.10
14	B	Sodium citrate	0.20
15	B	Purified water	83.65
		Characteristics	Transparent and soft gel
		Sensory characteristics	Smooth and very light

Directions

1) Heat 1–5 to 50°C and dissolve. Disperse 6. (A)
2) Dilute 9 with 10. (D)
3) Dissolve 11–14 in purified water. (B)
4) Add A gradually into B while stirring. Dissolve completely to make a gel. Add 7, 8 and D, and add water. (B + A + C + D)

Specifications
pH: 5.9–6.9

Ingredient
*6: Lipidure-C (NOF corporation)

Prescription 5.49 Transparent wax for making lustrous hair

		Ingredient	% (100 g)
1	A	PEG-40 hydrogenated castol oil	16.00
2	A	Ceteth-40	16.00
3	A	PG	8.00
4	A	Hexyl decanol	4.00
5	A	Jojoba oil	0.20
6	A	Pentylene glycol	2.00
7	A	Phenoxyethanol	0.80
8	A	Steartrimonium chloride (70%)	2.00
9	B	Purified water	51.00

Directions

1) Weigh 1–5 and heat to 75°C. (A)
2) Heat purified water to 75°C. (B)
3) Pour into containers while it is still hot

5.5.5 *Hair Mist Liquids*

5.5.5.1 Damage-Care Hair Mists (Milky)

Most hair mist products are liquid, consisting of steartrimonium chloride and glycol, and do not contain oils.

Prescription 5.50 is a combination of amino modified silicone emulsion and a liquid emulsion of liquid oil and higher alcohol of cetostearyl alcohol emulsified by ceteth, PPG-ceteth and steartrimonium chloride. The oil and amino modified silicones are included for the purpose of repairing damaged hair.

Prescription 5.50 Damage-care hair mist (milky)

		Ingredient	% (100 g)
1	A	Cetostearyl alcohol	0.30
2	A	Ceteth-2	0.20
3	A	PPG-4 ceteth-10	1.50
4	A	Jojoba oil	0.50
5	A	Steartrimonium chloride (70%)	1.00
6	A	BG	2.00
7	A	Phenoxyethanol	0.50
8	A	Methyl paraben	0.10
9	B	Amodimethicone, steartrimonium chloride, cocodimonium chloride (*9)	2.00
10	B	Polyquaternium-64 (*4)	0.50
11	pH	50% Lactic acid or sodium lactate	
12		Purified water	to 100 g
		External appearance	Milky lotion
		Sensory characteristics	
		Texture	Smooth lubrication
		Tackiness	Slightly tacky
		Hair setting performance	None

Directions
1) Heat 1–8 to 80°C and homogenize. (A)
2) Heat purified water to 80°C. Add A and emulsify (350 rpm, 20 min). (A)
3) Cool to 45°C. Add 9 and 10. Adjust pH with 11, and add water. (A+B)

Specifications
pH: 4.5–6.0

Ingredients
*9: SM8904 Cosmetic emulsion (Dawcorning Toray Co., Ltd.)
*4: Lipidure-C (NOF corporation)

5.5.5.2 Hair Setting Mists

Hair setting polymers are included as in gels. Hair setting mists should be smooth, silky and not tacky and have hair styling performance.

Prescription 5.51 is a combination of steartrimonium chloride and hair setting polymers. The quantities of the constituents are kept low to control tackiness. The sensory characteristics of the glycols should be known in advance because they directly affect the feel of using the product (See Sect. 2.4, Chap. 2.).

Prescription 5.51 Hair setting mist (light)

		Ingredient	% (100 g)
1	A	PG	3.00
2	A	Methyl paraben	0.10
3	A	Steartrimonium chloride (70%)	0.50
4	A	PEG-60 hydrogenated castol oil	0.20
5	A	PVP	0.50
6	B	Sodium coco PG-dimonium chloride phosphate (*21)	0.50
7	B	*N*-Methacryloyl oxyethel *N,N'*-dimethyl ammonium-α-*N*-methyl carboxy betaine and Alkyl methacrylate copolymer (*25)	7.00
8	B	Glycerylamidoethyl methacrylate/stearyl methacrylate copolymer	0.30
9	A	Purified water	to 100 g
		External appearance	Transparent liquid
		Sensory characteristics	
		Hair setting performance	Weak
		Tackiness	None
		Hair unity	Good
		Texture	Light and lubricating at application. Silky finish

Directions
1) Heat 1–5 to 70°C, and homogenize. Add purified water to make a transparent liquid. (A)
2) Add 6 to 8, and homogenize. Add water. (A + B)

Specifications
pH: 6.0–7.0

Ingredients
*21: Arlasilk Phospholipid CDM (Croda Japan)
*25: Yukaformer R205S

5.5.5.3 Transparent Hard Hair Setting Mists

Prescription 5.52 is a hair setting mist for dispenser pumps. It is a modification of a hard hair spray in aerosol spray cans, which use propane as the propellant gas. Instead of propane, isododecane, and ethanol are used to accelerate drying. Isostearyl alcohol and diisopropyl adipate are combined to unite isododecane and ethanol, which are mutually incompatible and differ in polarity, into a single phase.

AMP-acrylates/diacetoneacrylamide copolymer, which is a hair setting polymer that can be mixed with cationic surfactants, is combined with anionic AMP-polyurethane-14, acrylates copolymer to achieve hard setting performance. The differences in characteristics by the combination of hair setting polymers and cationic surfactants should be noted (Tables 5.16 and 5.17). The inclusion of steartrimonium chloride gives an incomparable smooth texture not seen in conventional hard hair sprays.

Table 5.16 Hair mist (combinations of setting polymers and cationic surfactants)

Basic prescription	
Ingredient	Amount % (100 g)
Setting polymer	
1,3-Butylene glycol	3.00
Diethylhexyl succinate	3.00
Steartrimonium chloride (70%)	1.00
Purified water	24.00
Lactic acid	0.10
Ethanol	68.9
Total	100.00

5.5.5.4 Hair Lotions Not Containing Cationic Surfactants

Hair finishing lotions that do not contain cationic surfactants can be produced by combining moisturizing polymers such as *Tremella fuciformis* polysaccharide, sodium hyaluronate, hydroxypropyltrimonium hyaluronate, and sodium polyglutamate. Moisturizing hair lotions can be designed by combining PCA-Na, moisturizing polymers and anionic surfactants, such as TEA-cocoyl glutamate and sodium dilauramidoglutamide lysine.

Prescription 5.53 is a hair essence that gives a light finish. It consists of moisturizing polymers, (*T. fuciformis* polysaccharide, sodium hyaluronate, hydroxy-

Table 5.17 Differences in physical and sensory characteristics by hair setting polymer

Hair setting polymer	Amount of polymer	Ethanol	pH	Viscosity	Characteristics	Spread at application	Setting Strong 5>1 Weak	Hair unity United 5>1 Spread	Tackiness Strong 5>1 Weak	Texture
Octylacrylamide/acrylates/butylaminoethyl methacrylate copolymer(*14)	10.00	90.00	5.00	Transparent liquid		Slightly fastening	Rather weak, 2	Spread, 2	Tacky, 3	Moist and smooth
AMP-acrylates/diacetoneacrylamide copolymer AMPD(*15)	6.00	94.00	7.50	Transparent liquid		Not fastening	Rather united, 4	Rather united, 4	Weak, 2	Not smooth and dry
Polyurethane-14, AMP-Acrylates Copolymer(*9)	10.00	90.00	7.30	Transparent liquid	Forms complexes with cationins (becomes cloudy with little alcohol).	Not fastening and dry	Normal, 3	Rather united, 4	Strong, 4	Heavy, moist and smooth
PVP/VA(*1)	6.00	94.00	4.10	Transparent liquid		Not fastening	Rather weak, 2	Rather spread, 3	Tacky, 3	Light
Ethyl ester of PVM/MA copolymer(*4)	6.00	94.00	3.80	Transparent liquid	To be neutralized with cationic surfactants and not with alkali. Separates when diluted with water.	Heavy but not fastening	Very strong, 5	Highly united, 5	Very strong, 5	Highly lubricating

(continued)

Table 5.17 (continued)

Hair setting polymer	Amount of polymer	Ethanol	pH	Viscosity	Characteristics	Spread at application	Setting Strong 5 > 1 Weak	Hair unity United 5 > 1 Spread	Tackiness Strong 5 > 1 Weak	Texture
Octylacrylamide/acrylates/butylaminoethyl methacrylate copolymer(*7)	10.00	90.00	5.50	Transparent liquid	Forms complexes with cationins (becomes cloudy with little alcohol).	Not fastening	Strong, 4	Rather united, 4	Slight, 1	Smooth and lubricating
N-Methacryloyl oxyethel *N,N'*-dimethyl ammonium-α-*N*-methyl carboxy betaine and Alkyl methacrylate copolymer(*5)	10.00	90.00	6.70	Transparent liquid		Smooth and not fastening	Normal, 3	United, 4	Strong, 4	Light and slightly dry
Polyquaternium-11(*3)	15.00	85.00	5.70	Transparent viscous gel	Not suitable for hair mist due to high viscosity	Smooth and not fastening	Strong, 4	United, 3	Slight, 1	Light and rather smooth
Polyquaternium-16 (*17)	7.50	92.50	4.40	Transparent liquid		Smooth and not fastening	Normal, 3	United, 3	Not tacky, 0	Smooth and lubricating
Polyquaternium-55(*18)	30.00	70.00	4.20	Transparent viscous gel	Not suitable for hair mist due to high viscosity	Smooth and not fastening	Very strong, 5	Rather united, 3	Rather strong, 4	Moist, smooth and lubricating
Polyquaternium-49(*11)	6.00	94.00	5.50	Transparent liquid		Slightly dry, not fastening	Almost none, 0	Spread, 1	Weak, 2	Smooth, soft and lubricating

Prescription 5.52 Transparent hard mist (hair setting)

		Ingredient	% (100 g)
1	A	AMP-polyurethane-14,acrylates copolymer (*18)	4.00
2	A	AMP-acrylates/diacetoneacrylamide copolymer (*16)	9.00
3	A	Isostearyl alcohol	5.00
4	A	Diisopropyl adipate	2.00
5	A	Purified water	0.10
6	A	Steartrimonium chloride (70%)	0.30
7	B	Ethanol	79.40
8	B	Geraniol	0.10
9	B	Phenethyl alcohol	0.10
10	C	Isododecane	16.00
		External appearance	Transparent liquid
		Sensory characteristics	
		Texture	Smooth
		Tackiness	Tacky
		Hair setting performance	Strong

Directions
1) Heat 1–5 and dissolve. Add 6 and homogenize. (A)
2) Add B to A. (A+B)
3) Add C. (A+B+C)

Ingredients
*16: Plas cize L-53P (Goo Chemical Co., Ltd..)
*18: DynamX (AkzoNobel)

propyltrimonium hyaluronate, and sodium polyglutamate), PCA-Na, and anionic surfactants, such as TEA-cocoyl glutamate. It gives a moisturized touch different from those that use cationic surfactants such as steartrimonium chloride. The sensory characteristics vary depending on the hair moisturizing polymers used.

Prescription 5.53 Transparent hair lotion (toilet water)

		Ingredient	% (100 g)	% (100 g)	% (100 g)	% (100 g)
1	A	BG	4.00	4.00	4.00	4.00
2	A	Phenoxyethanol	0.80	0.80	0.80	0.80
3	A	PEG-60 hydrogenated castol oil	0.30	0.30	0.30	0.30
4	A	Methyl paraben	0.10	0.10	0.10	0.10
5	C	*Tremella fuciformis* polysaccharide (1% solution) (*26)	4.00	–	–	–
6	C	Sodium hyaluronate (1% solution)	–	4.00	–	–
7	C	Hydroxypropyltrimonium hyaluronate (1% solution) (*27)	–	–	4.00	–
8	C	Polyglutamic acid (1% solution) (*28)	–	–	–	4.00
9	C	PCA-Na	1.00	1.00	1.00	1.00

(continued)

Prescription 5.53 (continued)

		Ingredient	% (100 g)	% (100 g)	% (100 g)	% (100 g)
10	C	Sodium dilauramidoglutamide lysine (30%)	0.50	0.50	0.50	0.50
11	C	TEA-cocoyl glutamate (30%)	0.50	0.50	0.50	0.50
12	pH	Lactic acid	0.12	0.12	0.11	0.11
13	B	Sodium citrate	0.20	0.20	0.20	0.20
14	B	Purified water	88.48	88.48	88.49	88.49
		Sensory characteristics				
		At application	Slightly lubricating and adhesive	Moist and slightly slimy	Lubricating and moist	Moist and light lubrication
		Hair unity	Good	Good	Good	Good
		Moistness/silkiness	Moist	Moist	Moist	Moist
		Lubrication	Good	Rather good	Light lubrication	Light lubrication
		Tackiness	Not tacky	Not tacky	Not tacky	Not tacky
		Texture	Slightly dry and light lubrication	Smooth and lubricating	Smooth, soft and lubricating	Smooth and light lubrication

Directions
1) Heat 1–4, and homogenize. (A)
2) Dissolve 13 in 14. (B)
3) Add A to B. Add C, and homogenize. (B + A + C)
4) Adjust pH with 12, and add water,

Specifications
pH: 5.5–6.5

Ingredients
*26: Tremoist-TP (Nippon Fine Chemical Co., Ltd.)
*27: Hyaloveil (Kewpie Corporation)
*28: Bio-PGA Na Powder (Ichimaru Pharcos Co., Ltd.)

5.5.6 Hair Oils

Prescription 5.54 is hair finishing oil consisting of only silicones and oils that give lubricating and smooth textures. It contains highly polymerized dimethicone, dimethyl cyclopentasiloxane (cyclomethicone), and small amounts of vegetable oils.

The sensory characteristics of silicones are determined by the molecular weight of methyl polysiloxane. Those of 100 cs or lower give light oily feeling, those of 3500 cs or lower give strong oily touch, and those of higher cs give strong lubrication. High polymerization reduces oiliness and enhances lubrication. There is no difference in sensory characteristics between highly polymerized dimethiconol and highly polymerized dimethicone.

Prescription 5.55 is hair silicone oil that contains dimethicone/vinyl dimethicone cross polymer to slightly increase the viscosity. The viscosity of the product is determined by the amount of highly polymerized dimethicone, but dimethicone/vinyl dimethicone cross polymer may be added to achieve further thickening.

Highly polar oils can be combined into silicones by adding branched-chain ester, such as C14–18 alkyl ethylhexanoate and isotridecyl isononanoate, which blend well with silicones.

Prescription 5.54 Hair oil (highly lubricating with silicones)

	Ingredient	% (100 g)	% (100 g)	% (100 g)
1	Cyclomethicone dimethiconol (highly polymerized, 10%)	30.00	–	27.00
2	Cyclomethicone dimethicone (highly polymerized, 10%)	–	30.00	–
3	Dimethicone 100cs	–	–	3.00
4	Cyclomethicone	69.96	69.96	69.96
5	Tocopherol	0.01	0.01	0.01
6	*Simmondsia chinensis* (jojoba) seed oil	0.01	0.01	0.01
7	Squalane	0.01	0.01	0.01
8	Phytosteryl/octyldodecyl lauroyl glutamate	0.01	0.01	0.01
	Sensory characteristics			
	At application	Slippery lubrication	Slippery lubrication	Heavy lubrication

Homogenize 1–8

Prescription 5.55 Hair finisher (viscous product with silicone)

	Ingredient	% (100 g)
1	Cyclomethicone dimethicone (highly polymerized, 10%)	30.00
2	Cyclomethicone	55.80
3	Dimethicone/vinyl dimethicone cross polymer, cyclomethicone (*29)	5.00
4	Myristyl ethylhexanoate	2.00
5	Isotridecyl isononanoate	6.00
6	Squalane	0.20
7	Macadamia nut oil	0.20
8	Isostearyl alcohol	0.80
	Sensory characteristics	
	At application	Light lubrication and silky
	After drying	Good lubrication, favorable moistness

Homogenize 1–8

*29: KGS-15 (Shin-Etsu Chemical Co., Ltd.)

5.5.6.1 Hair Oil Not Containing Silicones

Hair oils that do not contain silicones can be produced by using isododecane instead of silicones. Various oily ingredients can be combined (Prescription 5.56). Combinations of oils produce liquids, and hydrogenated styrene/isoprene copolymer is used to thicken the liquid. The strong oily, moist and heavy texture of the hair oil given by the vegetable oils and fats and esters is modified into light and moist feel by adding isododecane.

Prescription 5.56 Oily hair conditioner (non-silicone)

	Ingredient	% (100 g)
1	Cholesterol	0.05
2	*Hordeum distichon* (barley) extract, *Santalum album* (sandalwood) extract, *Phellodendron amurense* bark extract	0.02
3	Phytosteryl/octyldodecyl lauroyl glutamate	0.02
4	Tocopherol	0.05
5	Squalane	0.20
6	*Macadamia ternifolia* seed oil	0.20
7	*Olea europaea* (olive) fruit oil	0.20
8	*Simmondsia chinensis* (jojoba) seed oil	0.20
9	Hydrogenated polydecene, hydrogenated styrene/isoprene copolymer	20.00
10	Propylene glycol isostearate	4.00
11	Cetyl ethylhexanoate	24.00
12	Isododecane	51.06

Direction
Homogenize 1–12

5.6 Cosmetics Mainly Consisting of Oils

There are cosmetics that contain little or no water, such as cleansing oils, lipsticks, and solid hair waxes. They are produced by combining carbohydrates, vegetable and animal oils and fats, waxes, esters, higher alcohols and nonionic surfactants and adding coloring agents, scenting agents and efficacious ingredients. Formulations are difficult to design, and techniques different from those for other cosmetics are required.

5.6.1 Characteristics and Techniques for Designing Formulations

Cosmetic products that mainly consist of oils may be solids, liquids, or pastes. All must be stable over a long period of time and against temperature changes. Products that do not contain water but are composed of only oils and surfactants are vulnerable to separation of liquid oil on the surface of the products. Major causes include:

1. Poor compatibility between the surfactant and oily components
2. Big changes in viscosity and property of the oily component by temperature and
3. Difference in melting point among oily ingredients

Separation of liquid oil can be prevented by improving the compatibility among ingredients by carefully investigating their polarities. For example, highly polar solid cetanol and nonpolar liquid paraffin cannot form a stable mixture but separate from each other.

The first step for improving the compatibility is to add an ester, which has a polarity intermediate of the two.

To achieve the product form (solid, liquid or paste), polar ingredients need to be selected based on the melting point. For producing liquid products, all ingredients must be liquid. Solid and paste products are produced by mixing liquid and solid components selected based on texture, viscosity, and ease of application.

When it is necessary to add surfactants, all or part of the surfactants must be nonionic surfactants. Surfactants of high HLB and anionic, cationic, and amphoteric surfactants destabilize the product and require a procedure of adding a small amount of water and/or ethanol, making a W/O emulsion and solubilizing the water and alcohol into the oily component to produce stable products.

5.6.2 Liquid Oil Cosmetics

Cleansing oils, self-emulsifying bath oils, and massage oils mainly consist of liquid paraffin and vegetable oils. They require nonionic surfactants for emulsification plus liquid esters and/or liquid higher alcohols that have intermediate polarity of the oils for homogenization, such as octyldodecanol, isostearyl alcohol, oleyl alcohol plus oleyl, unless the nonionic surfactant has an isostearyl group.

5.6.2.1 Cleansing Oils

Oil products for removing makeups consist of transparent liquid oils and nonionic surfactants. They are designed to dissolve the oily constituents of the foundation and lipstick and detach the coloring agents and inorganic powder components from the skin, which are then wiped or rinsed off.

The constituents include liquid paraffin, vegetable oils, esters, liquid higher alcohol, and nonionic surfactants.

The HLB of the nonionic surfactant should be decided based on the main oil constituent. Nonionic surfactant of relatively low HLB should be used for cleansing oils that mainly consist of liquid paraffin, and a high HLB should be used for those mainly consisting of liquid vegetable oils. Inappropriate combinations result in separation and cloudiness. To prevent separation and secure stability, higher alcohols and low-HLB nonionic surfactants are also needed to connect the major component and surfactants, which mutually differ in polarity. The amount of surfactant should be high in rinse-off cleansing oils.

Because they are also used around the eyes, the nonionic surfactant should be not irritating and must not be polyoxyethylene alkyl ethers. (Oleth-2, oleth-5 and oleth-10 included in makeup removers used around the eyes caused physical damage, and the Tokyo Metropolitan government investigated the case under the consumers' livelihood law. April 19, 2004, Bureau of Citizens and Cultural Affairs of the Metropolitan Government of Tokyo).

Prescription 5.57 mainly consists of PEG sorbitan tetraoleate and liquid paraffin.

Prescription 5.57 Cleansing oil

	Ingredient	% (100 g)	% (100 g)
1	Liquid paraffin	64.90	59.90
2	*Olea europaea* (olive) fruit oil	–	16.00
3	*Macadamia ternifolia* seed oil	10.00	0.50
4	*Simmondsia chinensis* (jojoba) seed oil	–	0.50
6	Octyldodecanol	10.00	–
7	Isostearyl alcohol	–	8.00
8	PEG-30 sorbitan tetraoleate	15.00	15.00
9	Natural Vitamin E	0.10	0.10
	Total	100.00	100.00
	External appearance	Yellow, transparent	Yellow, transparent

Direction
1) Homogenize 1–9

5.6.2.2 Bath Oils

The bath oil described below is those that self-emulsify and become cloudy when poured in bath.

It is recommended to use liquid paraffin as the main constituent because vegetable oils are prone to oxidization and smelling. Bath oils should be formulated so as to emulsify around 40°C because oils are emulsified with nonionic surfactant. As in cleansing oils, liquid esters, such as cetyl ethylhexanoate, and liquid higher alcohol, such as octyldodecanol, should be combined to help emulsification of liquid paraffin and vegetable oil. A prescription that uses sorbitan tetraoleate as the nonionic surfactant is exemplified below.

Prescription 5.58 is bath oil that becomes milky in bath. PEG sorbitan tetraoleate and sorbitan oleate are used to emulsify fluid paraffin and rice bran oil, which mutually differ in polarity. The HLB of the PEG sorbitan tetraoleate is adjusted to the polarity of the rice bran oil. Octyldodecanol and cetyl ethylhexanoate are added to improve the compatibility among the constituents and help them self-emulsify in water.

Prescription 5.58 Bath oil

	Ingredient	% (100 g)	% (100 g)
1	Fluid paraffin	65.00	–
2	*Oryza sativa* (Rice) bran oil	–	65.00
3	Cetyl ethylhexanoate	5.00	5.00
4	Octyldodecanol	10.00	10.00
5	Sorbitan oleate	–	6.00
6	Sorbeth-30 tetraoleate	20.00	–
7	Sorbeth-60 tetraoleate	–	14.00
	Total	100.00	100.00
	External appearance	Colorless, transparent	Colorless, transparent
	When dispersed in water	Becomes white and cloudy	Becomes white and cloudy

Direction
1) Homogenize 1–7

5.6.2.3 Massage Oils

Like bath oils, massage oils are transparent liquid consisting of liquid oils and nonionic surfactants.

The oil component should be smooth and lubricating to facilitate massaging. Compared to bath oils, reduced amounts of nonionic surfactants of lower HLB are combined because massage oils need not to self-emulsify. Because they may be not rinsed off but wiped off, the oily ingredients should be moisturizing and not tacky. Rinsed-off massage oils, which contain sorbitan oleate of low HLB as an emulsification agent, should be easy to rinse off.

Prescription 5.59 is oil that is smooth over a long period of time and oil that is stable against chronological oxidization and smelling.

Massage oils need not to self-emulsify and thus do not require a large amount of surfactants unlike bath and cleansing oils.

Prescription 5.59 Body massage oil

	Ingredient	% (100 g)
1	Mineral oil	75.00
2	*Olea europaea* (olive) fruit oil	4.00
3	Cetyl ethylhexanoate	4.00
4	Octyldodecanol	8.00
5	Sorbeth-30 tetraoleate	3.00
6	Sorbitan oleate	6.00

Direction
1) Homogenize 1–6

5.6.3 Solid and Stick-Type Cosmetics

Hard solid cosmetics consist of waxes, carbohydrates of high melting points, oils such as esters, and nonionic surfactants of low HLB. They are molded into solids, sticks, etc. or poured into containers. Nonionic surfactants are included to uniformly disperse iron oxides, coloring agents, pigments, and powdery constituents. Surfactants also help improve the spread, uniform application, unity with the skin, and retention on the skin of the products.

5.6.3.1 Concealers

Iron oxides, titanium oxides, talc, mica and other powdery constituents should be uniformly dispersed in concealers so that they can be spread thin and uniformly on the skin. To increase the spreading performance of waxes and carbohydrates of high melting points, the hardness of the product is adjusted by combining oils of various melting points. Isostearic acid is an effective nonionic surfactant for reducing smell and giving light texture.

Prescription 5.60 combines iron oxides as colorants into waxes, carbohydrates and nonionic surfactants of low HLB. The product should be easy to spread uniformly on the skin and have high skin covering performance by iron oxides and titanium dioxide. To minimize separation and viscosity changes by temperature, oils of various melting points and polarities are combined.

Prescription 5.60 Concealer

		Ingredient	% (100 g)
1	A	Candelilla wax	8.00
2	A	Beeswax	6.00
3	A	Behenyl alcohol	1.80
4	B	Polyglyceryl-2 isostearate	16.20
5	B	PEG-5 glyceryl stearate	5.40
6	B	Sorbitan palmitate	1.20
7	B	Shear butter	0.40
8	B	Petrolatum	7.20
9	B	Mineral oil	7.20
10	B	C14-18 Alkyl ethylhexanoate	22.22
11	B	Tocopherol	0.10
12	C	Black iron oxides	0.36
13	C	Yellow iron oxides	0.72
14	C	Red iron oxides	1.60
15	C	Titanium dioxide	16.20
16	C	Talc	5.40
		Total	100.00
		Hardness	Hard

(continued)

Prescription 5.60 (continued)

Ingredient	% (100 g)
Ease of application	Good
Staining	Good
External appearance	Solid

Directions
1) Weigh 1–11, heat to 80°C and dissolve. (A+B)
2) Add 12–16 into (A+B), and homogenize. (A+B+C)
3) Pour into containers while it is still hot, and cool

5.6.3.2 Stick-Type Cosmetics

Solid cosmetics molded into sticks mainly consist of carnauba wax, beeswax, behenyl alcohol, and surfactants of low HLB. Iron oxides are used as colorants in hair dye products. The spread and adhesion on the hair is controlled by combining oils of high melting points and nonionic surfactants.

Prescription 5.61 is a hair color stick. The lipstick-like dye is used to temporarily color the hair. It is applied on gray hair to make it inconspicuous and can be easily removed by shampooing.

Ingredients of high melting points are used in large amounts to mold the product into sticks. High-melting point oils of various polarities are combined to prevent thinning and separation of oils at high temperatures. It should be easy to spread and apply, give a light texture, and be not tacky.

Prescription 5.61 Hair color stick

		Ingredient	% (100 g)	% (100 g)
1	A	Carnauba wax	12.00	12.00
2	A	Microcrystalline wax	4.00	4.00
3	A	Beeswax	8.00	8.00
4	A	Behenyl alcohol	2.15	2.15
5	B	Polyglyceryl-2 isostearate	24.00	24.00
6	B	PEG-5 glyceryl stearate	4.00	4.00
7	B	Sorbitan palmitate	1.00	1.00
8	B	Mineral oil	8.00	8.00
9	B	Octyldodecyl myristate	16.00	16.00
10	B	Diphenyl dimethicone	6.00	6.00
11	B	Tocopherol	0.05	0.05
12	C	Black iron oxides	6.00	9.00
13	C	Red iron oxides	2.40	1.20
14	C	Yellow iron oxides	2.40	0.60
15	C	Talc	4.00	4.00
		Color	Brown	Black

(continued)

Prescription 5.61 (continued)

Ingredient	% (100 g)	% (100 g)
Hardness	Good	Good
Ease of application	Good	Good
Staining	Good	Good
External appearance	Solid	Solid

Directions
1) Weigh 1–11, heat to 80°C and dissolve. (A+B)
2) Add 12–15 into (A+B), and homogenize. (A+B+C)
3) Pour into containers while it is still hot, and cool

5.6.3.3 Fragranced Balm

Fragranced balms are hard creams of essential oils and/or perfume and are applied on a part of the body. The hardness is adjusted by combining ingredients of high melting points and liquid oils so that it is easy to apply. Oils of high melting points, such as cadelilla wax, beeswax, and behenyl alcohol, are used combined with nonionic surfactants of low HLB. The base has a component similar to that of concealers and stick-type cosmetics. High-melting point oils of various polarities are combined to prevent thinning and separation of oils at high temperatures (Prescription 5.62).

Prescription 5.62 Balm of essential oils (fragranced balm)

		Ingredient	% (100 g)
1	A	Candelilla wax	8.00
2	A	Beeswax	6.00
3	A	Shear butter	2.00
4	A	Behenyl alcohol	1.50
5	A	Polyglyceryl-2 isostearate	20.00
6	A	PEG-5 glyceryl stearate	5.00
7	A	Sorbitan stearate	1.50
8	A	PEG-60 hydrogenated castor oil	2.00
9	A	Octyldodecyl myristate	20.00
10	A	Tocopherol	0.10
11	A	Rice bran oil	32.30
12	B	Lavender oil	0.80
13	B	Eucalyptus oil	0.40
14	B	Rosemary oil	0.40

Directions
1) Weigh 1–11, heat to 80°C and dissolve. (A)
2) Add 12–14 into (A), and homogenize. (A+B)
3) Pour into containers while it is still hot, and cool

5.6.3.4 Solid Hair Wax

Prescription 5.63 is hair styling solid wax that contains polyethylene, which is carbohydrate of high melting point. The setting performance and sensory characteristics of the wax can be modified by changing the combination of polyethylene and esters.

In the prescription, water is added to form a W/O emulsion. Polyethylene and dimer dilinoleyl dimer dilionoleate are added as hair styling ingredients. The sensory characteristics are determined by the ester used.

Prescription 5.63 Hair styling clay wax

		Ingredient	% (100 g)	% (100 g)
1	A	Petrolatum	43.50	43.50
2	A	Kaolin	1.00	1.00
3	A	Titanium dioxide	0.50	0.50
4	A	Talc	5.00	5.00
5	B	Beeswax	19.00	19.00
6	B	Polyethylene	5.00	5.00
7	B	Dimer dilinoleyl dimer dilionoleate	5.00	5.00
8	B	Myristyl ethylhexanoate	5.00	–
9	B	Diethylhexyl succinate	–	5.00
10	B	PEG-6 oleate	2.00	2.00
11	B	PEG-60 hydrogenated castor oil	4.00	4.00
12	C	Purified water	10.00	10.00
		Hardness	Hard	Hard
		Tackiness	Rather tacky	Tacky
		Setting performance	Rather strong	Strong

Directions
1) Add 2–4 into 1 and homogenize. Heat to 80°C and dissolve. (A)
2) Add 5–11 into (A), dissolve, and add C. Homogenize, and pour into containers. (A+B+C)

5.6.3.5 Skin Protection Cream

Prescription 5.64 is a cream for protecting the skin along the face line from hair coloring and permanent wave agents. The paste-like cream, which mainly consists of petrolatum, covers the skin and blocks hair coloring and permanent wave agents from reaching the skin.

Carbohydrates of different melting points, esters, and cetanol are combined with surfactants of low HLB. The prescription is stable against high temperature and chronological separation. It can be uniformly applied on the skin thick and is soft and easy to wipe off.

Prescription 5.64 Skin protection cream

	Ingredient	% (100 g)	% (100 g)
1	Microcrystalline wax	16.00	6.00
2	Mineral oil	24.00	–
3	Petrolatum	–	36.00
4	Cetyl ethylhexanoate	36.70	34.70
5	Tocopherol	0.10	0.10
6	Cetanol	16.00	16.00
7	Sorbitan stearate	4.80	4.80
8	Polysorbate 80	0.80	0.80
9	Polysorbate 65	1.60	1.60
	Characteristics	Hard cream	Gel-like soft cream

Directions
1) Heat 1–7 to 80°C and dissolve. Add 8 and 9

5.7 Facial Toner

Facial toner should be formulated so as to moisturize, soften, firm up, brighten, and smoothen the skin and give a moisturized feeling. In quasi-drug products for whitening the skin and preventing wrinkles (anti-aging), derivatives of Vitamin C and ferment extracts are combined, respectively. Many facial toner products also contain ceramides, vitamins, sterols, plant extracts and other efficacious ingredients.

5.7.1 Sensory Characteristics of Facial Toner

The sensory characteristics of facial toner are mainly determined by the moisturizing ingredient used and its amount. Glycerin gives a rather strong moist feeling, and 1,3-butylene glycols makes the product rather light to use. Saccharides, such as trehalose, raffinose, and mannitol give a light and dry texture.

5.7.2 Safety of Ingredients

Facial toner requires special attention on the smell and skin irritation. Particularly, preservatives and oils must be carefully investigated by examining safety data and checking the safety in the product.

Although it is easily overlooked, ethanol becomes irrigating at high concentrations. Hydrolyzed proteins have peculiar smells and cannot be combined in large quantities.

5.7.3 *Properties and Constituents of Facial Toner*

Facial toner may be transparent liquid, transparent viscous liquid or milky lotion.

Oils are rarely combined in transparent liquid products. To include oils, large amounts of nonionic surfactants for solubilizing the oils and 1,3-butylene glycol or another glycol that has an OH group are needed. Polar oily ingredients, such as sterols and oily vitamins, can be solubilized by lowering the polarity of water by adding large amounts of nonionic surfactants of both high and low molar numbers and 1,3-butylene glycol.

Use of carboxy vinyl polymer and sodium acrylate is recommended for thickening transparent viscous liquid products. They little affect the sensory characteristics and make facial toner that is easy and smooth to apply. Oily ingredients make the product milky. Inclusion of a large amount of emulsifier results in heavy and moist textures. Nonionic surfactants that little affect the sensory characteristics should be used, such as PEG-60 hydrogenated castor oil.

5.7.3.1 Facial Toner with Solubilized Vitamins

Vitamins A and E, which are highly polar oils, can be solubilized, while minimizing the amount of surfactants, by combining PEG-60 castor oil of high HLB and a nonionic surfactant of low HLB such as PEG-7 glyceryl cocoate (Prescription 5.65).

Citric acid is also combined to solubilize magnesium ascorbate phosphate. The resultant pH becomes slightly alkaline (Prescription 5.66).

Prescription 5.65 Facial toner with solubilized vitamins A and E

		Ingredient	% (100 g)
1	A	PEG/PPG/polyputylene glycol-8/5/3 glycerin	4.00
2	A	Methyl paraben	0.10
3	A	Propyl paraben	0.05
4	A	Glycerin	5.00
5	A	PEG-60 hydrogenated castor oil	0.60
6	A	PEG-7 glyceryl cocoate	0.30
7	A	Tocopherol	0.10
8	A	Vitamin A	0.10
9	C	Sage leaf extract	0.10
10	C	*Rehmannia chinensis* extract	0.10
11	B	Dipotassium glycyrrhizate	0.10
12	B	Sodium lactate	0.20
13	B	Sodium citrate	0.20
14	B	Purified water	89.05

Directions

1) Dissolve 2 and 3 in 1 by heating. Add 4–8 and homogenize. (A)
2) Dissolve 11–13 in purified water. (B)
3) Mix A and B. (A + B)
4) Add 9 and 10, and water. (A + B + C)

Prescription 5.66 Facial toner with vitamin C included

		Ingredient	% (100 g)
1	A	BG	4.00
2	A	Pentylene glycol	1.20
3	A	Phenoxy ethanol	0.60
4	A	Glycerin	4.00
5	B	Magnesium ascorbate phosphate	1.00
6	B	Sodium citrate	0.50
7	C	*Gardenia florida* extract	0.30
8	C	*Sophora angustifolia* extract	0.30
9	C	Lactobacillus/pear juice ferment filtrate	0.30
10	C	Raffinose	0.50
11	B	Purified water	to 100 g
		pH	8.00

Directions
1) Homogenize 1–4. (A)
2) Dissolve 5 and 6 in purified water. (B)
3) Add A to B. Add 7–10 and water. (B+A+C)

5.7.3.2 Facial Toner Containing Ferment Extracts

For antiaging and preventing wrinkles, *Saccharomyces* barley seed ferment extract, *Lactococcus* ferment, and hydrolyzed yeast extract are included (Prescription 5.67).

Prescription 5.67 Toilet water with ferment extracts included

		Ingredient	% (100 g)
1	A	BG	5.00
2	A	Glycerin	5.00
3	A	Pentylene glycol	0.80
4	B	Sodium lactate	0.20
5	C	*Artemisia capillaris* flower extract, clove flower extract, etc. (*30)	1.00
6	C	Saccharomyces barley seed ferment extract	0.50
7	C	Lactococcus ferment	0.50
8	C	Hydrolyzed yeast extract	0.50
9	C	*Oenothera biennis* extract	0.10
10	C	Lactobacillus/grape juice ferment extract	0.10
11	pH	Lactic acid	
12	A	Purified water	to 100 g

Directions
1) Add 1, 2, and 3 in purified water. (A)
2) Dissolve 4 in A. (A+B)
3) Add 5–10 into (A+B). (A+B+C)
4) Adjust pH with 11. Add water

Specifications
pH: 4.5–5.5

Ingredient
*30: SY-Plantex (Sakamoto Yakuhin Kogyo)

Good textured facial toner containing moisturizing polymers.

Silky and smooth texture can be achieved by combining highly moisturizing polymers such as sodium polygamma-glutamate, *T. fuciformis* polysaccharide, and sodium hyaluronate as well as PCA-Na and sodium dilauramidoglutamide lysine (Prescription 5.68).

Prescription 5.68 Good textured toilet water containing moisturizing polymers

		Ingredient	% (100 g)	% (100 g)	% (100 g)
1	A	Carboxy vinyl polymer	0.40	0.40	0.40
2	B	Dipotassium glycyrrhizate	0.10	0.10	0.10
3	B	Betaine	0.50	0.50	0.50
4	B	Sodium citrate	0.20	0.20	0.20
5	B	Sodium lactate	0.40	0.40	0.40
6	C	BG	4.00	4.00	4.00
7	C	Glycerin	4.00	4.00	4.00
8	C	Carprylyl glycol	0.80	0.80	0.80
9	C	Ethylhexyl glycerin	0.20	0.20	0.20
10	D	Sodium polygamma-glutamate (1%) (*28)	10.00	–	–
11	D	*Tremella fuciformis* polysaccharide (1%) (*26)	–	10.00	–
12	D	Sodium hyaluronate (1%)	–	–	10.00
13	E	PCA-Na	2.00	2.00	2.00
14	E	Sodium dilauramidoglutamide lysine	0.30	0.30	0.30
15	E	Polyphosphorylcholine glycol acrylate (*31)	0.20	0.20	0.20
16	pH	10% Sodium hydroxide			
17	A	Purified water	to 100 g	to 100 g	to 100 g

Directions
1) Disperse 1 in purified water. (A)
2) Add 2–5 in A. (A+B)
3) Add 6–9 to (A+B). (A+B+C)
4) Add 10–12 to (A+B+C). Add 13, 14 and 15. (A+B+C+D+E)
5) Adjust pH, and add water

Specifications
pH: 6.0–7.0

Ingredients
*28: Bio-PGA Na Powder (Ichimaru Pharcos Co., Ltd.)
*26: Tremoist-TP (Nippon Fine Chemical Co., Ltd.)
*31: Lipidure-HM (NOF corporation)

5.8 Cosmetics not Containing Specific Ingredients

Prescriptions that do not contain preservatives in the positive list are exemplified in this section, including creams that do not contain silicones and shampoos that do not contain sodium laureth sulfate or sodium lauryl sulfate. The cosmetics are appealing to users as safe products free of preservatives, sodium laureth sulfate, and sodium lauryl sulfate.

5.8.1 Transparent Shampoo Consisting of Ingredients of Vegetable Origin

The transparent shampoo in Prescription 5.69 consists of ingredients of vegetable origin and uses ethylhexyl glycerin and grapefruit seed extract as preservatives. The pH is to be adjusted at 5.5–6.0.

The anionic surfactant is disodium cocoyl glutamate, and lauryl glucoside serves as the thickener. Cationic guar gum is used instead of polyquaternium-10. Dimethylamino propyl stearamide, which is a tertiary amine, is included to improve hair conditioning.

Prescription 5.69 Organic shampoo

		Ingredient	% (100 g)
1	A	Guar hydroxypropyl trimonium chloride	0.60
2	A	Glycerin	2.00
3	A	Betaine	0.50
4	A	Sodium citrate	0.40
5	B	Dimethylamino propyl stearamide	0.10
6	B	Disodium cocoyl glutamate (30%)	24.00
7	B	Cocamidopropyl betaine (30%)	12.00
8	B	Lauryl glucoside	4.00
9	B	Ethylhexyl glycerin	0.30
10	C	Grapefruit seed extract	0.50
11	C	PCA-Na	0.50
12	D	Orange oil	0.40
13	D	Grapefruit oil	0.10
14	D	Eucalyptus oil	0.05
15	E	Sage leaf extract	0.10
16	E	Lavender extract	0.10
17	E	Rosemary extract	0.10
18	pH	50% citric acid	
19	A	Purified water	54.25

Directions

1) Disperse 1 in 2. Add the dispersion into 19. Heat to 75°C. Dissolve to transparency. Dissolve 3 and 4. (A)
2) Add 5–9 into A and heat to 75°C. (A+B)
3) Cool (A+B) to 45°C. Add 10–17 to (A+B). (A+B+C+D+E)
4) Adjust pH with 18. Add water

Specifications

pH: 5.5–6.5

5.8.2 Pearl Shampoo Not Containing Sodium Laureth Sulfate

Prescription 5.70 uses *Artemisia capillaris* flower extract, clove flower extract, and phenoxy ethanol as preservatives. The pH is to be adjusted at 5.5–6.0.

Shampoos not containing sodium laureth sulfate are prone to thinning and destabilization. Ethylene glycol distearate and higher alcohols are used to improve high-temperature stability. The combination of the tertiary amines and anionic surfactants of sodium lauroyl alanine and sodium methyl cocoyl taurate gives a smooth texture. To keep stable viscosity at low and high temperatures and ensure chronological stability, two nonionic surfactants are combined as thickeners. The shampoo is smooth and moisturizing.

Prescription 5.70 Pearl shampoo not containing sodium laureth sulfate

		Ingredient	% (100 g)
1	A	Polyquaternium-10 (high viscosity, low amine)	0.50
2	B	Dimethylamino propyl stearamide	0.60
3	B	Cetanol	0.50
4	B	Oleyl alcohol	0.30
5	B	Ethylene glycol distearate	2.00
6	B	Cocamide MEA	3.00
7	B	Cocamide methyl MEA	1.00
8	C	Sodium lauroyl methylaminopropionate (30%)	12.00
9	C	Sodium methyl cocoyl taurate (30%)	24.00
10	C	Cocamidopropyl betaine (30%)	9.00
11	D	Phenoxy ethanol	0.70
12	D	Orange oil	0.40
13	D	Grapefruit oil	0.10
14	D	Lavender oil	0.05
15	D	*Artemisia capillaris* flower extract, clove flower extract, etc. (*32)	0.50
16	pH	50% citric acid	0.80
17	A	Purified water	45.90
		pH	5.80

Directions
1) Disperse 1 in purified water. Heat to 75°C. Dissolve to transparency. (A)
2) Add 2–7 into A, heat to 75°C, and dissolve to transparency. (A+B)
3) Add 8, 9 and 10. Cool to 45°C. (A+B+C)
4) Add 11–15 into (A+B+C). (A+B+C+D)
4) Adjust pH with 16. Add water

Ingredient
*32: SY-Plantex (Sakamoto Yakuhin Kogyo)

5.8.3 Shampoo Not Causing Buildup of Hair

Prescription 5.71 uses pentylene glycol and phenoxy ethanol as preservatives. Cationic guar gum and polyquaternium-7 are used instead of polyquaternium-10. It does not harden the hair even when used long but keeps the hair smooth and light.

Prescription 5.71 Shampoo not causing buildup of hair

		Ingredient	% (100 g)
1	A	Guar hydroxypropyl trimonium chloride	0.80
2	A	BG	3.00
3	A	Pentylene glycol	1.50
4	A	Sodium citrate	0.40
5	A	Betaine	0.50
6	B	Polyquaternium-7	1.00
7	B	Sodium lauroyl methylaminopropionate (30%)	24.00
8	B	Sodium methyl cocoyl taurate (30%)	8.00
9	B	Cocamidopropyl betaine (30%)	9.00
10	B	Cocamide methyl MEA	4.00
11	C	PEG-20 sorbitan cocoate	1.20
12	C	Lavender oil	0.08
13	C	Orange oil	0.40
14	C	Phenoxy ethanol	0.80
15	pH	50% citric acid	0.60
	A	Purified water	44.72
		pH	6.25

Directions

1) Disperse 1 in 2 and 3. Add the dispersion into purified water. Heat to 75°C. Dissolve. Add 4 and 5 and dissolve. (A)
2) Add 6–10 into A. (A + B)
3) Homogenize 11–14. (A + B + C)
4) Cool to 50°C. Add C. Adjust pH with 15, and add water

Specifications

pH: 6.0–6.5

5.8.4 Hair Conditioner Not Containing Specific Ingredients

The conditioner of Prescription 5.72 uses *A. capillaris* flower extract, clove flower extract, and phenoxy ethanol as preservatives. Dimethylamino propyl stearamide, behetrimonium chloride, and hydroxypropyl arginine C12–14 alkyl ether HCl help emulsification and improve the sensory characteristics. The oil content is increased to give a smooth and moisturizing feel.

Prescription 5.72 Hair conditioner

		Ingredient	% (100 g)	% (100 g)
1	A	Dimethylamino propyl stearamide	1.80	1.80
2	A	Behetrimonium chloride (80%)	0.80	–
3	A	Hydroxypropyl arginine lauryl/myristyl ether HCl	–	0.80
4	A	Glyceryl stearate	1.20	1.20
5	A	Cetanol	4.20	4.20
6	A	Cetyl ethylhexanoate	4.80	4.80
7	A	Macadamia nut oil	0.60	0.60
8	A	Octyldodecanol	0.60	0.60
9	B	Orange oil	0.70	0.70
10	B	*Artemisia capillaris* flower extract, clove flower extract, etc. (*32)	1.00	1.00
11	B	Phenoxy ethanol	0.60	0.60
12	C	Lactic acid	0.60	0.60
	C	Purified water	to 100 g	to 100 g

Specifications
pH: 3.0–5.0

Directions
1) Dissolve 1–8 by heating to 80°C. (A)
2) Add 12 to purified and heat to 80°C. (C)
3) Add A–C and emulsify. (C+A)
4) Cool to 45°C. Add 9–11 into (C+A). Add water. (C+A+B)

Ingredient
*32: SY-Plantex (Sakamoto Yakuhin Kogyo)

5.8.5 *Cream Using Food Additives as Surfactants*

Prescription 5.73 (a modification of Prescription 5.25) uses sodium isostearoyl lactate and sucrose stearate as surfactants.

Xanthan gum is added as a stabilizer. The oil content is increased to enhance smoothness and moistness. *A. capillaris* flower extract and clove flower extract can be used as preservatives because no stearic acid is combined and the pH can be adjusted below 6.

Prescription 5.73 Cream of food additive surfactant

		Ingredient	% (100 g)
1	A	Xanthan gum	0.20
2	A	Sucrose stearate (*4)	1.20
3	A	Glycerin	2.00
4	A	BG	2.00
5	B	Sodium isostearoyl lactate	0.60
6	B	Glyceryl stearate	2.40
7	B	Behenyl alcohol	0.80
8	B	Cetyl ethylhexanoate	3.00
9	B	Sesame oil	1.20

(continued)

Prescription 5.73 (continued)

		Ingredient	% (100 g)
10	B	Tocopherol	0.10
11	B	1,2-Pentandiol	0.80
12	B	Dimethicone 6cs	1.20
13	C	*Artemisia capillaris* flower extract, clove flower extract, etc. (*32)	0.50
14	pH	50% Lactic acid	to 100 g
		Purified water	84.00

Directions
1) Disperse 1 and 2 into 3 and 4, add into purified water, heat to 80°C and dissolve. (A)
2) Heat 5–12 to 80°C, and dissolve. (B)
3) Add B to A, and emulsify. (B + A)
4) Cool to 45°C. Add 13. (A + B + C)
5) Adjust pH with 14, and add water

Specifications
pH: 5.0–6.0

Ingredients
*4: SURFHOPE C-1816 (Mitsubishi Kagaku Foods)
Note: Sucrose stearate is difficult to disperse or dissolve in oil and is prone to form lumps
*32: SY-Plantex (Sakamoto Yakuhin Kogyo)

5.8.6 Gel Cream Not Containing Specific Ingredients

Prescription 5.74 is a light-textured cream that uses polymers. Carboxyvinyl polymer and xanthan gum are used as thickeners. The pH is around 7. Pentylene glycol and glyceryl caprate are used as preservatives.

Prescription 5.74 Gel cream

		Ingredient	% (100 g)
1	A	Carboxyvinyl polymer	0.80
2	B	Glycerin	4.00
3	B	Xanthan gum	0.10
4	C	Dipotassium glycyrrhizate	0.10
5	C	Betaine	0.50
6	C	Trehalose	0.10
7	D	BG	4.00
8	D	Pentylene glycol	1.20
9	D	Glyceryl caprate	0.90
10	D	PEG-60 hydrogenated castor oil	0.60
11	D	Squalane	1.20
12	D	Cetyl ethylhexanoate	0.40
13	D	Tocopherol	0.10
14	E	Sodium hyaluronate solution (1%)	2.00

(continued)

Prescription 5.74 (continued)

		Ingredient	% (100 g)
15	E	PCA-Na	0.50
16	pH	10% Potassium hydroxide	6.00
	A	Purified water	to 100 g

Directions
1) Disperse 1 uniformly in purified water while stirring. (A)
2) Disperse 3 in 2. (B)
3) Heat 7–14 to 70°C and homogenize. (D)
4) Add B to A, dissolve C, add D and E, and homogenize. (A + B + C + D + E)
5) Add 16 to neutralize and gelatinize. Add water

Specifications
pH: 6.0–8.0

Index

H. Iwata and K. Shimada, *Formulas, Ingredients and Production of Cosmetics: Technology of Skin- and Hair-Care Products in Japan*, DOI 10.1007/978-4-431-54061-8,